THE IMPENETRABLE FOREST

THE IMPENETRABLE FOREST

▼

Thor Hanson

Writer's Showcase
San Jose New York Lincoln Shanghai

The Impenetrable Forest

Writer's Showcase
an imprint of iUniverse.com, Inc.

For information address:
iUniverse.com, Inc.
5220 S 16th, Ste. 200
Lincoln, NE 68512
www.iuniverse.com

Cover Photograph: Copyright © 1993 Paul Joynson-Hicks
Used by permission

ISBN: 0-595-13018-6

Printed in the United States of America

For my parents—
the fisherman who taught me to persevere
and the artist who gave me words

CONTENTS

AUTHOR'S NOTE

In March of 1999, Uganda's Bwindi Impenetrable Forest gained international notoriety when eight tourists and a local warden were brutally massacred by exiled Rwandan Hutus. A major U.S. newsmagazine reported that "Africa's implacable hatreds" had claimed more innocent lives, playing on the misperceptions so common in our culture's view of sub-Saharan Africa. By portraying the killings as unknowable and in some way inevitable, they could simply be categorized as the latest tragedy from a region plagued by violence. Of course Uganda and its neighbors seem distant, foreign and complex, but we cannot afford to be so completely uninformed. The killers at Bwindi were the same Rwandans responsible for the genocide of up to a million of their Tutsi countrymen in 1994. If the international community had known more, empathized more and intervened then, those soldiers and militia would not still be armed and active today, terrorizing neighboring countries. The massacre at Bwindi demonstrates all too well that in the shrinking modern world, ignorance and inaction in the face of suffering will ultimately affect us all.

I offer this book in part to help fill that gap of knowledge, and hope that in addition to the passages on gorillas and natural history, the reader will take away stories and images that spark a greater understanding and curiosity for the people and cultures of Africa.

ACKNOWLEDGEMENTS

I am deeply indebted to the government and people of Uganda, and to the U.S. Peace Corps for providing me with the platform for this experience. I would also like to thank my long-suffering family and friends for putting up with me—both while I was away and while I was writing this book.

I think the author of any personal narrative owes everything to the people and places that make up their story. While I can never thank every person (or gorilla), let me acknowledge the friendship and support of these key characters: Liz Macfie, Kelleni Archabald, John Dubois, Rob Rothe, Dave Snedecker, Kathleen Henderson, Ted Hazard, Eunice Kang, Ed, Steve, Monica, Erwin, Alice, Jenny, Erin, Scott, Kelly, J, Steve, Rose Ssebatindira, the Ntale family, Commander J. J., Ignatius Achoka, Blasio Bbyekwaso, Phenny Gongo, Hope Nsiime, Tibesigwa Gongo, Betunga William, Alfred Twinomujuni, Medad Tmugabirwe, Caleb Tusiime, Levi Rwahamuhanda, Ephraim Akampurira, Agaba Philman, Abel Muhwezi, Magezi Richard, Stephen Mugisha, Prunari Rukundema, Charles Kyomukama, James Tibamanya, James Mishana, Yosamu Baleberaho, George Gachero, Benjamin Bayende, Gaston Kanyamanza, Benon, Daudi, Christopher, Stanley, Dominico, and all the good folks from Bizenga to Kanyashande.

I'm grateful for the tireless efforts of my agents Mary Alice Kier and Anna Cottle, and to Richard Nelson for his insightful editorial comments on the manuscript.

Also, I would like to acknowledge Houghton Mifflin Company for permission to quote from the works of Jane Goodall[1] and Dian Fossey,[2] Alfred A. Knopf, a division of Random House, for the use of a quote from Michael Crichton's *Congo,*[3] Harper Collins for the Robert Hass translation of Basho,[4] and Lonely Planet Publications for material from their 1997 *Guide to East Africa.*[5] Finally, many thanks to Paul Joynson-Hicks for the use of his marvelous Bwindi gorilla photo as a cover graphic.

1 from *In The Shadow of Man* by Jane Goodall. Copyriht © 1971 by Hugo and Jane van Lawick-Goodall. Reprinted by permission of Houghton Mifflin Co. All rights reserved.
2 from *Gorillas in the Mist* by Dian Fossey. Copyright © 1983 by Dian Fossey. Reprinted by permission of Houghton Mifflin Co. All rights reserved.
3 Copyright © 1980 by Michael Crichton. Reprinted by permission of Alfred A. Knopf, a division of Random House. All rights reserved.
4 from *The Essential Haiku*, edited by Robert Hass. Copyright © 1994. Reprinted by permission of Harper Collins. All rights reserved.
5 Copyright © 1997 by Lonely Planet. Reprinted by permission of Lonely Planet Publications. All rights reserved.

PROLOGUE

"I speak of Africa
and golden joys."

—William Shakespeare
King Henry IV, 1597

From my back window the forest rose in a wall of green, every possible shade, as if the whole spectrum were made from one color. Pale branches and tree trunks tangled through the foliage in an incomprehensible lattice, like veins of wood in a vast emerald sea. The stillness seemed ancient, a heavy, tropical pall broken only by the water-drop cries of a coucal, or the soft chatter of waxbills. At dusk, the heat and birdsong gave way to cricket-rasp, a deafening metallic backdrop to nights of mist and moonlight. I slept with the shutters ajar, half-alert, listening for another kind of sound, haunting and unmistakable: the staccato chest beats and deep coughing barks of a mountain gorilla family, my closest rainforest neighbors, and the subject of my work in Africa.

I joined the Peace Corps in the summer of 1993, committed to the idea of two years abroad with only the vaguest of job descriptions: 'Natural Resource Management in Uganda.' I'd turned down a similar post in The Gambia after a trip to the library yielded only four articles about that tiny,

riverbank country—all from medical journals about malaria. The literature on Uganda seemed comparatively vast, and while many stories
focused on the despotic reign of Idi Amin, they also mentioned the
Mountains of the Moon, The Great Rift Valley, and the Source of the
Nile, romantic features of a landscape that Winston Churchill dubbed
"the pearl of Africa." The nature of my assignment, however, remained a
mystery until I reached the country and sat for my last, pre-service inter-
view. While the other volunteers answered questions about tree-planting,
teaching, and agricultural skills, my interrogation had a different focus:

"Do you work well in isolation?"

*"How would you feel about living seventeen kilometers from the nearest
market?"*

*"What can you tell us about your experience with field studies and large
animals?"*

"Can you picture yourself living in a remote tropical rainforest?"

I answered every query with feigned confidence, and it paid off when
our supervisor, Rose Ssebatindira, called everyone together to announce
the final site placements. We gathered around a large map and she read
down the list of volunteers, reaching my name last. "Thor..." she said,
pointing vaguely at a roadless area in the far southwest corner of the country: "We're sending you to the middle of nowhere."

If the Peace Corps had held a job lottery, then I surely hit the 'Triple
Seven.' Rose's "middle of nowhere" turned out to be an isolated patch of
jungle called Bwindi Impenetrable Forest, home to elephants,
chimpanzees, monkeys, woodland antelope, and nearly half the world's
population of mountain gorillas. For two years I worked with Uganda
National Parks and The International Gorilla Conservation Program
(IGCP), helping develop tourism in the newly-formed Bwindi
Impenetrable National Park. As supervisor of in-forest activities, I spent
my days habituating wild gorillas, training local guides, constructing a
network of trails, and struggling to find a sense of place in the tiny village
where I made my home.

The project in Bwindi introduced gorilla tourism as an economic incentive for conservation, an integral piece of the ongoing effort to save the species. As a self-admitted eco-nerd, this assignment was a dream job and I fully expected mountain gorillas to be the focus of my life in Uganda, and the subject of any writing that might come of it. But as early as our training period, when I lived with a local family near Kampala, I found myself equally captivated by the warmth and generosity of the Ugandan people, and by the country's complex culture and history. I came to realize that one cannot discuss the future of gorillas meaningfully without placing them in the social and political context of modern Africa. Succesful conservation efforts depend upon reliable human frameworks; parks and wildlife populations are only as stable as the systems that support them. Uganda's mountain gorillas exist as part of the inseparable weave of people and landscape, and there will not be hope for either until there is hope for both.

So while a good portion of this narrative deals with mountain gorillas, the book shouldn't be mistaken for an in-depth analysis of the species. My observations were made over long periods of daily contact, but our program was designed for tourism, not research. For a detailed scientific examination of gorillas, I refer the reader to the unparalleled works of Dian Fossey and George Schaller. Additionally, I've included a select list of references for people interested in learning more about Uganda's rich history and cultural heritage, and there is a glossary to explain any words or phrases in dialect.

The following chapters and events fall roughly in chronological order. I begin by establishing cultural context the way it was established for me, through total immersion in the life of a Ugandan family. The story continues through two years in the Impenetrable Forest, and ends, as any African journey should end, on the palm-shaded beaches of Zanzibar. The characters encountered along the way are real people, and I'm deeply grateful to all of them for allowing me to tell their stories, as well as my own.

CHAPTER I

WARAGI DREAMS

▼

*"Waganda, as I found subsequently, are not in the habit of
remaining incurious before a stranger. Hosts of questions
were fired off at me about my health, my journey, its aim,
Zanzibar, Europe and its people, the seas and the heavens, sun,
moon and stars, angels and devils, doctors, priests, craftsmen…"*

—Henry Morton Stanley reaches Uganda, *1875*

"Do you drink?"

The question caught me off guard and I hesitated, searching his somber
eyes for a hint of expectation. I hardly knew the man. Did *he* drink? A 'yes'
answer might offend him, but a 'no' sounded too puritanical.

"Uh,…sometimes?" I answered cautiously.

Jackpot. His eyes lit up and a broad smile creased his bearded face. For
the first time in our several hours together, Tom Ntale looked excited,
even eager.

"Wait here," he said and rushed out of the room, a curtain of tiny yellow fly beads rattling and swaying across the doorway in his wake.

I breathed a sigh of relief and found myself momentarily alone in the
house. Susan had disappeared next door with Aunt Florence, and Sam was
at the spring fetching water. My back ached from sitting stiffly all afternoon

and my face felt bent and sore from smiling. I took the opportunity to stand and stretch, and examine the Ntale family sitting room in detail.

If Elvis and my grandmother had ever conspired to decorate a tree fort, this would have been it. Two overstuffed red velour chairs matched my own crimson seat, crowded around a narrow coffee table and facing a large velvety sofa. Crocheted pink and lime antimacassars lay draped across each piece of furniture, and the panes of a small, glass-fronted credenza were pinned across with beaded lace. Canary yellow walls peeked out from behind a kaleidoscopic clutter of photos and advertisements, cut from old magazines: Lux brand facial soap, Pepsi cola, House of Manji Tea Biscuits, and a bug spray called Doom, guaranteed to "kill all *dudus.*" There was a framed photograph of a yawning lion, a collection of grainy black and white family portraits, and a large poster of the Pope. Brightly patterned sisal mats covered the concrete floor, a final touch in this haphazard collision of African and Colonial English design. Behind me, I noticed two shelves in a high corner filled with rows of empty bottles—Scotch, bourbon, gin—displayed like a trophy case. *Uh oh.*

Beads rattled suddenly and Tom was back, gripping two glasses and a tall, ominous-looking bottle of clear liquid.

"Local whiskey," he told me, sitting down to pour. "We make it from bananas."

Tom handed me a full glass and we made toasts to America and Africa before taking hearty gulps of what appeared to be jet fuel. A hot fire poker burned down the inside of my throat and into my stomach. I looked away, eyes watering, trying not to cough or choke out loud.

"Ahh," Tom said, staring at his half-empty glass with satisfaction. "How do you see it?"

"Good," I croaked as the flames spread into my limbs, and my face flushed red. "Tastes…bananas," I managed to add, with what I hoped was a winning smile.

"Yes," he seemed genuinely pleased. "We ferment banana juice, then distil it. Sometimes twice. This one is called *waragi.*"

When the fire began to fade, I did finally taste a hint of banana, as if someone may have dipped an old peel briefly into the vat of spirits. Only then was I sure that my host hadn't mistakenly served me a glass of kerosene. I've heard that good moonshine burns with a blue flame, the sign of healthy, intoxicating ethyl alcohol. Booze that burns yellow is a methyl-based solvent, where the hangover symptoms sometimes include blindness or raving insanity. I took another sip of the *waragi*. Definitely yellow flame.

Tom clapped his hands suddenly. "Photographs!" he exclaimed, and began rummaging through a pile of scrapbooks and old newspapers beneath the coffee table. "You must see our snaps!" He pulled out an armful of albums and loose pictures, smiled, and topped up our glasses.

Yellow flame or no, the *waragi* brought with it a welcome release from the day's awkward formality. Tom and I had exhausted our supply of polite greetings early on, sitting uncomfortably silent while twenty other cross-cultural pairs slowly formed up around us. It was the first full day Peace Corps training for Uganda's 1993 volunteer group, and families from all around the small town of Kajansi had come to our training center to collect us for a ten week stay in their homes. When every wide-eyed American was paired with their respective hosts, we gathered for a short speech from the training director—thanks to the Ugandans and 'go team' encouragement for the dazed volunteers.

I listened numbly, watching a flock of Ross's turacos hop from branch to branch in the tree above us like huge emerald ravens. They called out in hoarse croaks, pecking at clusters of tiny blue fruit with their broad, clownish bills. The warm air smelled faintly of earth and blossoms, with the acrid tinge of burning grass, and wood smoke from a thousand cook-fires. Drums and singing drifted up from the valley below, and in the distance, over the farms and rooftops, Lake Victoria stretched away south to the horizon, a deep blue sea capping white in the breeze.

It was beautiful, exotic, and definitely a long way from Kansas. Simple jet-lag could have told me that, let alone moving in to the village home of an African family. The speeches continued and I glanced sideways at my new 'father,' a proud, serious-looking man of middle years, with thick grey hair, a high forehead and wide-spaced, solemn eyes. He seemed more stern than the other parents, and I wondered briefly if he thought he'd gotten a lousy volunteer. I was short, sunburned, and young enough to be his son; and I was sweating like a boxer.

We loaded our luggage into Peace Corps vans and drove away from the training center, entering a storybook African landscape of lush green hillsides sprinkled with mud-brick houses. Rusted tin roofs and fields of red-berried coffee trees wavered in the heat, and great crowds of people walked before us on the narrow, rutted roads. Brightly dressed women balanced hoes and heavy bundles on their heads, while men sped past on bicycles, or pushed wooden wheelbarrows overflowing with strange goods and produce: pineapples, passion fruit, long poles of sugarcane, charcoal stuffed in burlap sacks, goat hides, huge clay pots, and teetering piles of firewood. I drank in the scene like a voyeur glimpsing some forbidden world of vibrant colors, reeling noise and bewildering vitality, while throngs of children ran alongside the van, laughing, waving, and shouting *"Abazungu, abazungu! How are you fine!"*

We passed through the center of Kajansi town, a dense cluster of dusty painted buildings and shops fronting the busy Entebbe highway. Houses were packed around the trading centers and scattered among the fields with no discernable plan. Expensive homes with high-walled compounds stood directly next to unkempt mud-brick shacks. Cows, chickens and goats ran wild through the town, and any spare patch of land supported crops: maize, cassava, papaya, passion fruit vines and the tall, broad-leaved stems of bananas, everywhere bananas.

The van lurched down a narrow side street and stopped in front of Tom's house, a fine brick and tile structure shaded by jack-fruit trees and sandwiched between a grove of banana stems and two small potato fields. He shouldered my backpack as the van honked and pulled away, and we turned to walk up the path.

"Sam!" he called out, "Susan!"

A rangy, narrow-shouldered youth in blue Adidas shorts careened around the side of the house and dropped to his knees, eyes downcast. Tom handed him my bag without an introduction, as the front door opened and a broad-faced woman in a bright orange dress hurried out into the sunlight. She too dropped to the ground before us and looked away, mumbling a greeting in Luganda.

Tom answered her, then said in a loud, stately voice, "Susan, this is our American guest, Thor." The letters r and l are interchangeable in most Ugandan dialects, so my name came out something like *Tall*. Susan murmured a new greeting, ending in Tall, and held out her hand in my direction. I reached over to shake it and she rose gracefully to her feet with a swish of satiny fabric, then disappeared back into the house.

"And that one is Sam," Tom added, as the boy hefted my luggage and followed Susan inside. I waved to him and his wary face was transformed by a wide, white-toothed grin.

"Hooo!" a voice called out from behind us. "You have brought our visitor!" We turned to find a large, heavyset woman advancing across the lawn from the neighboring house. She wore a dress even fancier than Susan's, brilliant green and purple with a wide sash, and shoulders that puffed up in tiny wings.

"This is Florence," Tom told me, "my sister. She stays in the house just there."

"Yes," Florence said, huffing a bit from exertion, "Sister and neighbor both." She bunched her skirt in one hand and I was afraid she too was going to kneel, but she stopped with a half curtsey, averted her eyes, and muttered a quick Luganda greeting before stepping forward to shake my hand with a warm smile.

"You will call me Aunt Florence," she stated firmly. "Already, you are a part of the family."

Florence ushered us both into the sitting room where we took our places side by side in the big red chairs, and proceeded to hold court for a seemingly endless procession of friends and relations. People stopped by from all over the neighborhood to get a look at Tom's new *muzungu*, the Luganda word for any non-African visitor.

Each arrival involved a complex, multi-lingual series of greetings, hand-shakes, and genuflecting children, with a fresh round of beverages from Susan. She carried in trays of Sprite and Fanta, always kneeling to serve the male guests. When she came around to me I took a bottle of Sprite and smiled warmly, but couldn't help thinking, *Please get up. It's just a soda.*

I held my tongue, however, trying to keep my mind open wider than the proverbial skies of Africa. I'd been told in advance to expect this ritual. Women and children in the Baganda tribe would typically drop to the ground as a formal greeting to men or elders. White visitors, men and women alike, might also be shown that symbol of respect, particularly in the more traditional families. As Susan backed away I made a mental note to find out how long such formalities might last. If I was really going to live with her for the next few months, it would be nice to make eye con-tact once in a while.

We began each visit with several minutes of smiling silence. Sometimes the group would talk to each other quietly, or ask Tom a question in Luganda, but their gaze never wandered from my face. I stared back, glassy-eyed with a blank grin, feeling the passage of time. Two years began to seem much longer than it had when I filled out the Peace Corps appli-cation. I heard the word *muzungu* a lot, and found myself wondering vaguely what they were saying while they appraised me:

'He sure is white, Tom.'

'Yes, yes. The man is very white.'

'From America?'

'Yes. And white.'

Invariably, someone would call me out of my daze with a carefully phrased question, "By the way, how do you find Uganda?"

"Very good," I would say, and, "You have a beautiful country."

"Thank you, please." They answered, and then, "How is Clinton?"

"The President?"

"Yes, yes. Bill Clinton." They pronounced the last name as two distinct words: '*Clinn-Tun.*'

"Well, just fine. He's just fine, I think."

"O.K. You are very welcome."

With that, the floodgate would open to an interrogation about my family, my home town, what crops we grew, airplanes, London, the journey from America. Slowly by slowly, the afternoon passed in a sea of dark faces and stilted English.

Seventy years as a British Protectorate showed itself in more ways than Tom's strange interior decorating. Names too bespoke a bizarre cultural juxtaposition: Nakibinge Matilda, Emma Magezi, David Rwandigito, Emmanuel Kayiwa, and Millie Nassolo. Pop-American culture had made its mark as well, with perhaps even more peculiar results. I met an insurance salesman named Commander J. J., a spice merchant called Splash Bobson, and later, a retinue of others: Happy James, Night Justice, Peace Marley, and a friendly young carrot farmer from Kabale, Lucky Dick.

As darkness settled in, the stream of visitors quickly tapered off and Tom and I found ourselves alone. He asked questions about America and we talked about the Peace Corps, but the conversation was stiff and awkward until it finally brought us around to the topic of banana whiskey. Tom's sudden change in demeanor may have been a product of alcohol, but it seemed to have less to do with intoxication, than with simple presence of drink. Be it *waragi,* banana beer, or pineapple wine, Tom always relaxed visibly when all the people around him had full glasses, and in the months ahead, we would have some of our best conversations at the local pub.

"This one is my brother's third child, at the baptism," he said, pointing to a faded snapshot. "I am the one here, behind, with the choir." We were half-way through the second photo album, and I was beginning to feel I'd recognize most of Tom's family on sight—at least if they were wearing graduation gowns, wedding dresses, or choir robes. Austere, formal poses

in a potato field dominated the layout, making the whole family look like
hard-luck Dakota farmers in a documentary on the Homestead Act.

"And here is Elvis."

"What?" For a wild instant I thought my interior decorator theory had
been right on the mark.

"My eldest son, Elvis Ntale," Tom clarified. "He is studying at
Makerere University. You see the books, there."

"Ah, yes, Elvis. You named him for the singer?"

"Yes, wonderful American music!" He clapped his hands and hummed
something unrecognizable. "I also love Dolly Parton, and Kenny Rogers."

I told Tom that the first record I ever owned was Elvis's Golden Greats.
He said he knew it and we each named our favorite songs, amazed and
gladdened that music would cross a cultural gulf to be the first bond in
our friendship.

We continued looking through the snapshots, a ritual for visitors in any
Ugandan household. Few people own cameras, but 'snaps' are incredibly
popular, and families often hire a photographer for special occasions. Later,
my own camera would garner me far more party invitations than my con-
versational wit, and I was thankful I didn't get stationed in eastern Uganda,
where circumcision ceremonies are the photographic subject of choice.

By the time we finished three picture books, *waragi* and lingering jet
lag were tilting the room at strange angles, and my eyelids sank into a
heavy, drunken squint. But Tom was just getting started.

"We will watch the television," he announced. "For you it will be just
like the States." He left the room shouting "Susan!" and returned momentarily carrying a large, surprisingly modern color TV.

"Tonight there should be enough power," Tom informed me as he
plugged the set into an outlet behind the couch. A single dam where the
river Nile tumbles out of Lake Victoria generates the bulk of Uganda's
electrical power. Unfortunately, that bulk is barely enough to supply
Kampala, let alone the outlying towns. Whole sections of the power grid

are shut down on a regular schedule, and even when the power is on, it's often not strong enough to run most appliances.

Still, having appliances to run was very unusual, even in the city. With electricity, a cassette player and a TV, the Ntale family ranked solidly in the upper half of Uganda's small middle class. Water still came in buckets from the spring, and a car was out of the question, but in a country where most people survived hand to mouth on subsistence agriculture, Tom lived in luxury. He worked as an accountant for the Ministry of Transportation and Public Works, taking draft payroll checks and vouchers from his office in Entebbe to the government bank in downtown Kampala.

"I am a *senior* accountant," he told me once with a laugh. "That means I don't have to do anything."

Florence and Susan followed Tom into the room and helped themselves to cups of *waragi,* while Sam rolled out a straw mat and sat cross-legged on the floor. He was the houseboy, I learned, the son of a distant cousin.

"From Mbarara district," Tom explained. "If you get a kid from the neighborhood they refuse to work hard, and then just run back home."

With the aerial twisted up against the ceiling, Tom fiddled with dials, and muttered under his breath, searching for the signal of Uganda's one, government-run TV station.

Suddenly an animated urban landscape swam into view, and an eerily familiar voice swelled out of the crackling speaker, "Hey, hey, hey, it's Faaaat Albert!"

I gaped at the screen.

"Ah, Fat Albert," Tom said with satisfaction, "One of your good American blacks."

I couldn't hide my disbelief, but I don't think anyone noticed. It was probably the same expression I'd had pasted across my face all day. As the evening wore on I found myself glancing around the room, stunned by the sheer oddity of the situation: drunk on fiery banana booze, watching politically incorrect cartoons from the 1970s in the sitting room of my new Ugandan family. I wanted desperately to communicate this with someone, to roll my eyes and wink at the hidden camera. But the Ntales didn't find the scene at all unusual. Susan still hadn't looked at me, Sam and Aunt Florence were glued to the screen, and whenever I turned to Tom, he only smiled and refilled my glass. Just when I thought I'd reached the pinnacle of cross-cultural whiplash, Uganda TV rolled into the second feature: Big Time Wrestling.

This, apparently, was everyone's favorite. Florence clapped her hands and shouted "Hoo!" every time someone took a full body-blow or got flung through the ropes. Even Susan was laughing, and Sam looked on happily with his sheepish, toothy grin. Tom wanted me to demonstrate some moves, and was actually sliding the coffee table aside to make room, when I finally convinced him that not all Americans knew how to wrestle.

"Is it?" he asked, looking incredulous and disappointed. My fledgling grasp of Ugandan English had already identified this phrase as equivalent to the American "*Really?*" or even closer, "*No shit?!*"

"These people must have very special training," I told him, improvising, as he glanced from me back to the sweaty men on the screen. I wasn't sure I could stand up too well at that point, let alone try to grapple my host into a full-nelson. "They're professionals."

This seemed to mollify him, and when a government report on coffee prices replaced the World Wrestling Federation, we finally moved into the adjoining room for dinner. It was past ten o'clock, an average time for supper in the Baganda tribe. Some families didn't dine until close to midnight, and for many volunteers, growing used to a strange new diet was far less challenging than actually staying awake to eat it.

Florence excused herself and disappeared next door, while I took a seat at the table with Tom and Susan. Sam kneeled beside me with a pitcher of warm water and I rinsed my hands, catching the run-off in a shallow bucket. When everyone had washed, Susan wordlessly handed Sam a bowl of food and he crouched down to eat on the floor at our feet. I was glad that Susan at least was eating with Tom and me. In some families, only the men have table privileges.

Tom mumbled a quiet grace and began dishing up heaps of steaming yellow paste from a bowl in the center of the table.

"*Matoke*," he identified. "The green bananas, steamed." There was rice too, and peeled sweet potatoes, beans, peanut sauce, some kind of pale, chalky root, and a bitter vegetable stew. I piled food onto my plate until I had some of everything, then paused, glancing casually around the table for cutlery.

Tom spotted my look, and shook his head. "Here, we use the Ugandan spoon," he said, holding up his hand.

"Oh, no problem," I assured him, immediately digging in to the mass of food. Unfortunately, for the drunk *muzungu* raised on knife and fork, nimbly transferring grains of soggy rice and peanut soup from plate to mouth was a problem. In fact, it was out of the question. A battleground of food shrapnel soon surrounded my plate and littered the floor beneath me. The meal progressed in silence, and I began to worry that my food mess constituted some unspeakable *faux pas*.

As if he'd read my mind, Tom stopped chewing and said, "In Ankole, Toro—some places they eat and talk. But not the Baganda. We are serious. When we eat, we eat."

And eat we did. Tom methodically devoured three huge servings, and any inroads I made into my own plate were quickly filled with another sweet potato or a ladle of beans. *Matoke,* the pasty banana substance, seemed to have good absorption powers and I ate more, hoping it would have a retroactive effect on the *waragi* sloshing around in my stomach.

"But you haven't eaten anything," Tom said, dismayed, when I turned down a fourth helping of beans. I'd weighed the potential consequences of refusing more food or becoming physically ill at the table, and decided to go with the former. "*Muzungus* don't know how to eat," Tom concluded with a shake of his head. "I have heard about this."

We sat in silence for a moment while Sam and Susan cleared the table, then he turned to me questioningly, "More television?"

"No thank you, Tom. I think I need to sleep." It seemed best to retire before Tom found another bottle of *waragi*.

"Yes," he said. "We will all sleep."

He led me down the corridor and pushed open the door to one of three rooms. With the kids away at school, I would have a bedroom to myself for the duration of my stay.

"For you we found a special bulb," he announced, groping for the switch. Suddenly a lurid, whorehouse glow flooded out into the hallway, and I peered inside to see a single, blood-red light bulb dangling from the ceiling of my new room.

"Perfect," I lied.

"You will be fine here, I think."

"Yes, yes. Thank you Tom," I assured him, and called into the shadows down the hall, "Thanks for everything, Susan. Goodnight."

"No, no," Tom interjected. "You will speak Luganda. For Susan you say *Sula bulungi, Nyabo.*'"

"*Sooben, lunger...*" I replied.

"Good, good," Tom chuckled and handed me something shiny. "The key to your room. Always keep it locked. There are thieves."

With that he moved off down the hall and I shut the door on my first day with the Ntales. In the glimmer of my special red lantern, I quickly surveyed the room. A loose sheet of orange and green linoleum covered part of the floor, and on it sat my backpack, a small wooden chest, a footstool, and, most importantly, a sturdy wooden bed. I staggered over and collapsed into a slumber that bordered on coma, and if I dreamed of home, the vision was gone by morning.

Chapter II

The Gold Dust Twins

▼

"I myself consider myself the most powerful figure
in the world and that is why I don't let any superpower
in the world control me."... "God is on my side.
Even the most powerful witchcraft cannot hurt me."

—His Excellency President for Life Field Marshal
Al Hadji Dr. Idi Amin Dada, VC, DSO, MC,
Lord of all the Beasts of the Earth and Fishes of
the Sea and Conqueror of the British Empire in
Africa in General and Uganda in Particular, *1975*

"We must take you to the police."

I paused with a bite of cold yams halfway to my mouth. "The police?"

"Yes, and the Resistance Council. We will go to both of them this afternoon when you return."

I chewed the yams, swallowed, and took a sip of tea, wondering silently what aspect of my behavior warranted calling in the authorities. Things, I thought, had been going fairly well. From our initial conversations about music and wrestling, Tom and I had developed an easy rapport, particularly after several evening tours of the local *waragi* bars. I'd learned to say 'good morning, sir,' in Luganda, and to sing a song called, 'I'm so very happy.' A small negotiation secured me a regular-colored light bulb for my bedroom, and I'd even mastered the one-handed rice and gravy scoop. When I'd taken ill with fever and had to spend a night at the training center, Tom had hiked there twice to visit, genuinely worried. I was still waiting to sustain a conversation with Susan, but she'd stopped kneeling every time I greeted her, and I took this as a good sign.

In fact, life with the Ntales had settled quickly into a steady routine and I was amazed at how quickly the wildly foreign can transform itself to normalcy. I woke at dawn every morning with a quiet knock at my door and three words from Susan, "water is ready." Bathing outside, I could hear the town come awake around me: roosters wailing at the dusty-rose light of sunrise, radios blaring Zairian pop-music, neighbors calling out greetings across the fields, and the patter of bare feet, as children ran to the spring for water.

After my yams and tea ("Always with tea," Tom told me. "If you take yams alone they will strike you mute."), I set off for the forty minute hike to training with David Snedecker, a West Virginian forester and my closest Peace Corps neighbor. Dave was only a few years older than me, but his height, thinning red hair and steel-rimmed spectacles led many Ugandans to think he was my father. Later, when my own hair grew long, they would take me for his wife. This didn't bother Dave. He had a fine-tuned sense of the bizarre, and once talked of building 'The Great Wall of Uganda' to stimulate tourism in his village. When asked by a Peace Corps trainer what he missed most from home, he answered simply: "Stilts." His humor, and his willingness to accompany Tom and me on back-street pub crawls, had led to a fast friendship.

Those first, overwhelmingly alien weeks in Uganda tied all the volunteers together, forming bonds between disparate people who would never associate under 'normal' circumstances: pot-heads and born-again Christians, poets and welders, Rush Limbaugh fans and lesbians. In a group that ranged from a twenty-two year old merchant marine to a seventy year old schoolteacher, bridging the cross-cultural gaps within our own American microcosm was sometimes just as great a leap as reaching out to Ugandans.

Our hilltop classrooms overlooked rolling hills and a patchwork of small *shambas*, or family farms, descending to the green, papyrus-lined shores of Lake Victoria. We learned basic language skills from among Uganda's forty dialects, as well as such Peace Corps essentials as 'how to light a charcoal stove,' or 'cooking a balanced meal with local ingredients.' I spent the time between classes with another fast friend, Rob Rothe, a caustic New Yorker who shared my enthusiasm for off-beat humor and ornithology. Occasionally, the two could be combined. On the first night at his homestay, Rob shared his bedroom with a family of chickens who mistook the whiteness of his skin for the pale glow of dawn. The rooster crowed directly into Rob's face for hours, and he returned the next evening dreading another sleepless night. But when the family gathered for dinner, his homestay father smiled and grandly pulled the lid from the stewpot, "Your chicken is here."

Usually we confined our birdwatching to the training center porch, sitting with binoculars glued to our eyes, and picking out new species from the constantly shifting flocks. Brown parrots clamored in the treetops, while dozens of white-throated bee eaters darted and swerved, bat-like, over the open fields. We saw long crested eagles, black kites, scarlet-breasted sunbirds, and two resident kingfishers, shining blue, perched silently in the shade like sentinels.

Walking home along the footpaths and red-dust roads was a circus parade of shouted greetings and laughter. People shook their heads and stared with looks of incredulous amusement. In the slow-to-change world of a Ugandan village, our passage was a daily spectacle whose novelty never faded. Throngs of children followed behind us in their matching green and blue school uniforms. The bravest among them would run along side, touch our hands, and dash back to their friends, screaming in time to the constant sing-song chorus: "Bye *Muzungu*! *Abazungu* Byee! How are you?"

Sometimes I stopped in the market, a warren of stalls stacked high with fresh fruits, produce and fish. Susan kept a small *duka* there, a simple shop where she sold beans, soap, maize flour and other household items to supplement her part time wage at the local clay works. We didn't talk much, but she would pull a small bench from behind the counter and we would sit together in the shade of her papyrus mat awning, watching the people watch us.

More often, I would run into Tom's spies—kids he posted in the street to interrupt my homeward route. "Uncle Tom wants to see you in the back!" they would cry, tugging my hand and leading me through the alleyways to Annette's place, the local pub where Tom met his friends every evening for drinks and conversation. The 'bar' consisted of two wooden benches and a low table in the dirt courtyard beside Annette's home. Scores of such establishments dotted the back streets of Kajansi, but Tom referred to Annette's as his 'club.' While ostensibly open to all, each bar had its own loyal clientele, and unaccompanied newcomers were often excluded, or driven off with verbal abuse.

Tom chose to drink at Annette's because, in addition to banana booze, she brewed the town's best *munanasi*, a delicious pineapple wine that quickly replaced *waragi* as the sane consumer's local beverage of choice. There were evenings, however, when heady drinks and bizarre dialogue appealed far less than a good night's sleep, and I tried skulking home unnoticed, or paying off the scouts—"Here's a thousand shillings, kid; you never saw me." But there was no such thing as an unobtrusive *muzungu* in Kajansi, and I rarely made it home without stopping by to greet Tom and the *munanasi* regulars: D.K., Vincent, Joseph, Chris, John Kennedy, and Sanyosenyo. We sat together in the dusty yard at sunset, sipping cups of wine and discussing their favorite topics, from world politics, to AIDS, to country music and, for reasons I never discerned, the city of Chicago.

I'd only been there once, a two-day layover on my way to Uganda. But I was American, and I'd seen "The Blues Brothers." This made me the local authority.

"They say that in Chicago there are more of your American blacks than all the people in Uganda. Is it true?"

"Well, I don't think it's that many, but—"

"And trains running through the sky?"

"Yes, the transportation is good, but I haven't—"

"By the way, how do you find Uganda?"

In all of this I could find no behavior screaming out for my arrest, so Tom's sudden decision to visit the police came as a surprise. When I asked the Ugandans on our training staff, they told me that some neighborhoods liked to have a record of all visitors to the area. It stemmed from a village custom where strangers were required to make a blood bond with the local chief. They said not to worry. And then they wished me luck.

Historically, good luck and Ugandan police stations haven't exactly gone hand in hand. As former Amin Cabinet Minister Henry Kyemba put it, "In Uganda, people who leave rooms escorted by policemen seldom come back." Police and soldiers arrested, tortured, and executed hundreds of thousands of people during the 1970s and early 1980s. Blame for the decades of chaos falls largely on two men, Milton Obote and Idi Amin Dada. Known as the 'Gold Dust Twins' for their early involvement in arms and bullion smuggling, Obote and Amin led Uganda to ruin as two of the most notorious despots in post-Colonial Africa.

Both rulers relied on the army to stay in power, and under the guise of 'internal security,' government soldiers operated unchecked by any laws save their own whims, or the caprice of their commanders. Looting and robbery were daily occurrences throughout the country, and unspeakable murders became commonplace as the army developed brutal killing methods to save on ammunition and instill terror in the populace. Prisoners were crushed by the dozens under moving tanks, burned alive, or forced to bludgeon one another to death with sledge hammers. No one was safe from persecution. In one famous case, Amin had the Anglican archbishop and two cabinet ministers dragged away from a public event and shot, then staged a ludicrous car accident to explain their deaths.

"We would not go outside the house for a week," Tom told me about the worst years of Amin's regime. "People disappeared every day."

So many vanished, in fact, that a lucrative new industry sprang up in Kampala, body finding. Entrepreneurs and off-duty soldiers would comb the prisons and mass grave sites, hired by families to find the remains of murdered friends and loved ones. The fees varied, depending on the status of the victim, from several hundred dollars to more than two thousand for high government officials and important political prisoners.

Memories of those years still haunt the Ugandan collective conscience, manifesting themselves in a great concern for security. In all the cities and major towns, thick iron grates cover every window, and many homes lie within high-walled compounds, guarded at night by dogs or watchmen. Communities contribute to train and equip local security forces, men and women from the area who volunteer to patrol the streets at night.

In Kajansi, people rarely walked anywhere outside of their immediate neighborhood after dark, and never alone. Returning from Annette's place with Tom, we always moved as a group and went far out of the way to escort any lone individuals to their houses. Ugandans traditionally give their visitors a 'push,' accompanying them for the first part of their journey home, and saying farewell somewhere along the path. This practice has simply been expanded for the hours after dark, ensuring that people are ushered home safely, all the way to their front door.

Once inside, families shuttered their homes tightly for the night. The doors at Tom's house were bolted, padlocked, and barred with thick ironwood timbers. Even so, we locked our individual bedrooms, and forgetting to do so was cause for serious reprimand. As a final precaution, a sort of home security ace-in-the-hole, the Ntales kept a rabid-looking, wild-eyed dog named Fox. I never saw more of the animal than a frenzied blur of teeth and drool lunging at the slats of his wooden cage, but Tom told me that Fox was something like a German Shepherd. He also told me I should never try to touch him. At night, the dog was set free to attack anyone or anything that intruded into the yard. "Fox is loose," Tom would say as he bolted the back door, and none of us would leave the house until dawn.

Discipline in the armed forces had improved dramatically since Yoweri Musevini's government took over in 1986, and public safety was no longer considered a major problem outside the troubled northern regions. The Ntale family's security measures seemed impregnable to me, and I thought it was surely a case of overkill until I was awakened late one night by a cacophony of high-pitched, ululating screams. Scrambling out of my mosquito net I clung to the barred window frame and peered outside. The banshee wails were definitely coming from a neighboring house, but I could see nothing clearly in the dim light of a sickle moon. Had someone died? Was there a fire? Or was this something completely normal, a tradition I had yet to learn about?

I heard Tom in the corridor and went out to find him unbolting the back door. He was carrying a rusty-looking revolver and he looked at me sternly, "Bandits have attacked the neighbors. Stay inside."

No problem. I returned to my room and had no trouble remembering to lock the door. I didn't know what kind of burglars attacked a well-secured, occupied home, and I didn't want to find out. But Tom was gone for over an hour, and as the noise gradually died down, I wondered if I shouldn't be doing something to help. I almost went next door to investigate, until the voice of reason pointed out that the sudden appearance of a frightened *muzungu* in pajamas probably wouldn't have a positive, calming effect on the situation.

The next morning he told me how the thieves had used a tree trunk to batter down the front door, and quickly made off with everything of value in the sitting room. The family's alarm cries had driven them off, and the neighborhood watch gave chase, but the half-dozen bandits, armed with spears, bows and arrows and a gun, had disappeared through the fields.

"Probably retrenched soldiers," he said, stirring sugar into his tea. "They will never catch them." The robbery fueled village gossip for weeks. Most people agreed with Tom, but there was another popular theory involving a disgruntled brother-in-law from Jinja. Everyone concurred that whoever it was would never be caught, and as far as I know they're still looking.

I later described the scene to my language tutor and learned a new verb, *kuteerateera*, 'to beat an alarm.' The original meaning had to do with pounding out rhythms on huge drums as a warning of danger to neighboring villages. Wailing screams were the modern equivalent, an effective way to alert the community and rally support.

"It's good when you hear *kuteerateera*," he told me. During the bad times, when government soldiers abducted people in the night, families didn't bother crying out because they knew their neighbors would be too afraid to help.

Amin's flamboyance and complete disregard for international opinion cemented Uganda's reputation as the center of despotism and human rights violations in Africa. His reign of terror during the 1970s is viewed as the darkest period of Uganda's history, but equal atrocities under Obote, and intervening periods of anarchy, prolonged the country's suffering. From the mid-1960s until the current government took power in 1986, Ugandans lived in a virtual police state, with no hope of economic or social stability. As industry and government services collapsed from mismanagement and corruption, many of the country's best-educated people either fled into exile or returned to the villages, relying on subsistence agriculture and waiting for the next coup to bring a new regime to power. Any review of those chaotic decades reveals the world of Ugandan politics as a constant struggle for control among myriad ethnic and religious factions. Successive governments fought only to keep themselves in power, preventing the country from ever developing a strong sense of national unity.

Over forty distinct ethnic groups live within the boundaries of modern day Uganda, a land area roughly the size of Oregon. Many tribes had developed highly organized kingdoms before The Berlin Conference of 1885 split Africa arbitrarily among the European powers. Long-held enmities and prejudices still ran strong and deep in the new Uganda, particularly between the Nilotic pastoralists in the north, who held sway in the army, and the Bantu-speaking peoples of the south and west, who had gained greater economic and political influence under British rule. The Colonial government exacerbated these rivalries by giving greater autonomy and influence to the powerful Baganda tribe and their traditional king, the *Kabaka*. The British also favored Protestant converts over Catholics and Moslems, creating a new and powerful religious division among the people.

As Independence approached in 1962, newly-formed political parties quickly divided the population along old lines. Candidates campaigning for Uganda's first elections rallied support among their home tribe, while the retiring Colonials used their influence to ensure a pro-British outcome. Finally, a coalition of Protestant and Bugandan parties won the majority of seats in Parliament, placing the new country in the hands of Prime Minister Milton Obote, with the Kabaka, Edward Mutesa, in the largely ceremonial position of president.

In spite of underlying political tensions, Uganda embarked upon self-government as one of the continent's most promising young nations. The country boasted excellent roads and infrastructure, as well as a growing industrial base, verdant agricultural land, and among the best social services in the sub-Saharan region. Makerere University in Kampala was regarded as the finest institute of higher learning in East Africa, and several prominent African leaders studied there, including Julius Nyerere of neighboring Tanzania. The breathtaking scenery and wildlife of "the pearl of Africa" attracted thousands of tourists to Ugandan game parks every year, and the international community held high hopes for the fledgling nation.

Ugandans remember the first years of independence with nostalgia and a touch of regret. "You should have been here in the '60s," Aunt Florence told me once. We were sipping *munanasi* on her porch while the sun set in a burgundy stain over Kajansi. "The economy was good and people had money to spend," she went on, and shook her head. "We were jolly; jolly indeed. You would have loved it."

The days of hope and jollity were short lived. Omens of conflict surfaced as early as the first Independence Day celebrations in Kampala, where crowds of Protestants waved Obote banners and the Bagandans cheered only their *Kabaka*. Factions who had united with autonomy from Britain as their only common goal began vying for power, and the coalition government soon showed signs of strain. Within three years, communication between Obote, Mutesa, and army commander Shaban Opolot had broken down completely, as each accused the other of assassination plots and planned coup d'etats. Finally, in 1966, Obote seized control of the government with a series of sweeping reforms. He arrested five cabinet ministers suspected of plotting against him, suspended the constitution, and ordered an attack on the *Kabaka's* palace with the help of his old smuggling partner in the armed forces, Col. Idi Amin. *Kabaka* Mutesa fled into exile, Opolot was arrested, and Uganda's brief career as a democratic nation came to an abrupt end.

Over the next four years, Obote consolidated his power with measures that set the precedent for two decades of totalitarian rule in Uganda. He rewrote the constitution, giving himself absolute authority in the role of President. Declaring a state of emergency in the Baganda homeland, and later the whole country, allowed him to suspend civil liberties and detain hundreds of opposition leaders without trial. Political parties were banned and all elections postponed indefinitely. He used political appointments and army promotions to control different factions within the government, and rampant corruption crippled the economy, as those in favor took full advantage of their positions. With his popularity waning throughout the country, Obote had to rely more and more on Amin to maintain control by force. But just as Obote had learned to manipulate the strings of government, Amin had increased his own influence and power as commander of the armed forces. The two began disagreeing openly in late 1970, and it came as no surprise when Amin staged a coup the following year, toppling Obote's regime and bringing the country finally under the direct control of the military.

Although promising that "free and fair elections will soon be held," Amin wasted no time entrenching himself and his allies in positions of power. As crowds filled the streets of Kampala to cheer Obote's downfall, the executions had already begun. Hundreds of soldiers and officers from Obote's Luo tribe were purged from the army, while prominent politicians and business leaders simply disappeared. Western countries who had backed Amin's rise to power began withdrawing their support, as tales of his eccentricities and violent inhumanity grew. Acting on instructions he claimed to receive in a dream (but more likely mimicked from his good friend Momar Ghadaffi's treatment of expatriate Italians in Libya), Amin expelled Uganda's entire Asian community in 1972. He redistributed their extensive business holdings among his political and military cronies, and an already weakened economy fell into ruins. Production in key areas from sugar and soap to coffee, cooking oil and cotton plummeted by as much as 91 percent over the next several years. Isolated from the international community, and threatened by growing internal opposition, Amin ruled the country by decree and force of arms alone. Daily flights from Europe brought plane loads of whiskey and expensive clothes to ensure loyalty among his troops, and they responded, sending thousands of suspected enemies, including foreigners and diplomats, into prison cells and unmarked graves.

I thought about this dark history as Tom and I approached the Kajansi police station, a small cement block structure at the edge of town, set back from the highway and shaded by two huge jacaranda trees. The building looked innocuous enough, but I wondered what kind of residual terror haunted its dull grey walls. Three officers in pressed khakis lounged on metal chairs by the roadside. We greeted them as we passed, and I noticed an AK-47 balanced casually against the stool between their feet. They gave us a bored wave, and one of them said something to Tom about the '*muzungu.*' Everybody laughed and so did I, falling back on a principle rule for travel anywhere in Africa: a policeman's jokes are always funny.

Inside, we waited on a wooden bench, watching a young taxi driver lodge some kind of vociferous complaint with a stone-faced officer behind the counter. He gesticulated wildly while he talked, pausing only to peer into an open ledger where the policeman took occasional notes with careful strokes of his pencil. The room was a cold, square box of scarred concrete, unpainted and smelling faintly of urine. Dim light filtered through window slits near the ceiling, and the atmosphere was oppressive and hopeless, exactly how one would picture a Ugandan police station. A heavily barred, three-quarter sized doorway in the back wall led further into the small building, and I sincerely hoped I would never find out what was on the other side.

The officer mumbled something without looking up and the driver nodded, suddenly humble, as if surprised by his earlier tirade. He pulled at the brim of his hat, a black baseball cap proclaiming "Bob Marley Lives," and backed out of the room, leaving me and Tom in a silence broken only by the slow scratching of the policeman's pencil.

Tom had hardly uttered a word all afternoon, and I suddenly wondered how he felt about our little visit to the police. Having lived through the reigns of Amin and Obote, he had certainly lost friends to the death squads, and police stations probably held far worse connotations for him than they did for me.

Finally, the desk officer beckoned us forward, and we stood in front of him like guilty schoolboys while Tom explained the situation. The officer listened, glanced at me with disinterest and pulled a different ledger book from the unruly pile on the desk behind him.

"Where are you from?" he asked in perfectly accented English, and I tried not to look startled. I told him I was from the United States.

"Which part?"

"Washington," I answered. I had long ago given up on explaining the difference between my home in the Pacific Northwest, and Clinton's home in the capital, but the officer surprised me again, becoming the only person in Uganda ever to ask me "The state, or D.C.?"

He then took down my name, and Tom's, and the location of Tom's house, entering everything slowly into his book. With that, he waved us away and I found myself suddenly outside again, blinking in the late afternoon sunshine. The whole process had taken less than ten minutes, and I had the impression that the policeman, far from being surly and threatening, was probably one of the best educated, most interesting Ugandans I'd met.

Still, I felt a distinct sense of anticlimax as we walked up the hill towards Annette's place. I was relieved, certainly, but also disappointed, as if the whole experience had failed to live up to its full dramatic potential. Ugandans must have had a similar reaction in 1979, when they finally got rid of Idi Amin only to see the same anarchy, dictatorship and civil war continue unabated for another seven years.

As the 1970s drew to a close, Amin found himself leading a shattered country with a restless, increasingly disillusioned army. To keep the troops occupied and rally popular support, he tried to annex Tanzania's sparsely populated Kagera region. When his forces swept southward, Ugandan exiles and rebel groups leapt at the opportunity to join the Tanzanians in repelling Amin's attack. The disheartened Ugandan army retreated steadily, and within half a year, Amin had fled to Libya and victorious Tanzanian forces swept into Kampala, handing over power to a coalition of more than twenty different Ugandan rebel factions.

The Ugandan National Liberation Front, as the union was known, had shared only one common goal: ridding the country of Idi Amin. With Amin gone, their cooperation dissolved quickly into a political struggle between the different groups, each with its own vision of how the country should be run. The Popular Front for the Liberation of Uganda (PFLU), The Uganda Patriotic Movement (UPM), The Front for National Salvation (FRONASA), The Relief Education Training Uganda Refugees Now (RETURN), The Save Uganda Movement (SUM), The Uganda Nationalist Movement (UNM),...the list goes on. Confusing enough in acronym form, the reconciliation of so many competing interests proved hopeless. Two presidents, Yusuf Lule and Godfery Bianisa, served less than six months each before a rigged election in 1980 paved the way for the return of Milton Obote.

Obote's second term in office met with immediate armed resistance. Former defense minister Yoweri Musevini went 'back to the bush,' with troops already experienced from their struggle against Amin. He was joined by Yusuf Lule to form the National Resistance Movement (NRM), and they began a grueling, five year guerilla war to bring down Obote's government. As the campaign raged on, Obote's countermeasures grew more desperate and brutal.

"At least with Amin, he killed only his enemies," Tom told me once. "Obote would kill anyone."

Government troops slaughtered whole villages in the Luwero region northwest of Kampala, a stronghold of NRM support. In less than five years, over 300,000 died, mainly innocent civilians and suspected opponents of the regime. But the NRM forces gradually gained ground, and by 1985, Musevini controlled most of Western Uganda.

A groundswell of discontent rose through the ranks of the government army, and for the second time, Milton Obote was overthrown by his top military commanders. General Tito Okello took over the office of president and immediately called for peace talks with the NRM and other warring factions. Several smaller groups joined his coalition, but a treaty with the NRM fell through, and fighting escalated along the front lines west of Kampala.

In January of 1986 the National Resistance Movement emerged victorious and installed Yoweri Musevini as the country's eighth president. Although he came to power by force, Musevini vowed that his presidency would mark a fundamental change in the political landscape, and Uganda embarked on a decade of steady recovery.

Musevini's government succeeded by maintaining a delicate balance between the country's major political factions. Instead of execution, he relied on negotiation as a governing tool, and his cabinet included many rivals and former enemies. When I arrived in 1993, Uganda's news media freely printed opposition viewpoints, and the country's human rights record had climbed from nightmarish depths to become one of the best in the region. Rapid economic growth and continued government stability gave rise to an atmosphere of tentative hope, and President Musevini was regarded as a hero by many Ugandans. Although civil unrest still marred the northern districts, and critics argued for a return to multi-party politics, the lively political debates I heard among Tom's friends all agreed on one point: the NRM government was good for Uganda.

"Without Musevini we would still be fighting," a *munanasi* regular told me once with solemn, bloodshot eyes, and everyone at the bar nodded their assent.

Life for the average citizen was steadily improving, but the habits learned in twenty years of violent unrest still dictated the rhythms of our daily routine. At ten o'clock sharp we locked ourselves into the house, and Fox was set loose to patrol the yard. I didn't begrudge Tom his elaborate safety precautions, particularly after the neighbor's robbery, but it did have drawbacks. Late one night I woke again, not to the cries of *kuteerateera*, but to face a far more personally threatening security problem. Clenched and sweating under my mosquito net, I cursed the combination of copious pineapple wine before bedtime, and a lack of indoor plumbing. With the doors tightly barred and a wild, slavering dog prowling the compound, a midnight latrine run was out of the question. I got up and paced in tight circles around the room, about to learn another critical lesson of the Peace Corps experience: 'how to fashion a bedpan from local materials.'

I pawed through everything in my backpack: walkman—no; sandals—no; flashlight—no; aspirin bottle?...not up to the task at hand. Time was growing desperately short when my search uncovered the only possible solution.

As part of our anti-malaria precautions, the Peace Corps equipped each volunteer with a vial of potent insecticide. A mosquito net soaked in the stuff was guaranteed to kill any bug in the room, and I'd seen a single drop of it literally melt the legs off a three-inch cockroach. While the effects on a person inside the mosquito net were never determined, we were warned to keep the concentrated solution from touching our skin, and every net-washing kit came equipped with a pair of large, clear plastic dish gloves. Calling on the cold-blooded tactics I'd learned in a third-grade water balloon fight, I easily filled one of the gloves and sealed it with a rubber band. Escapees in a Ugandan prison break never knew such relief, and I slept soundly for the rest of the night.

The next morning, however, I faced an even more challenging task: subtly disposing of the evidence. At first light I snuck outside with my makeshift bedpan and headed for the pit latrine, but even at dawn, 'inconspicuous *muzungu*.' is a contradiction in terms. I immediately met a group of neighborhood kids returning with water from the spring. They dropped their yellow jerry cans to the ground with a slosh, and crowded around me, giggling and chattering in Luganda.

"*Musuzi mutea*," I greeted them, nonchalantly hiding the glove behind my back. But it was no use; they'd seen exactly what I was carrying, and for that moment I needed no translator to understand the conversation: "Hey, look at that *muzungu*—he's got pee in a glove!! Har, har, har!"

I scuttled away towards the outhouse with my sloshing parcel, thinking in that instant of mortified clarity: *it's no wonder people stare at me in this country. I'm running around with a plastic glove full of urine.* I may not have built any cultural bridges that day, but desperate situations call for ridiculous innovation and I kept glove number two standing-by on the bedstand, primed and ready. With luck, it would be the worst Ugandan security emergency I'd ever have to face.

Chapter III

The Kabaka's Fence

▼————————

*"It was in those days a fine native town, extending from the
Kabaka's enclosure on the top of the hill Mengo fully a mile to
the north, east and west, with well-kept roads fenced on each side
with elephant grass and tidy courtyards to each enclosure...."*

—John Roscoe

The Soul of Central Africa, 1922

"*Jjuko*, we are going!" Tom's voice echoed down the corridor as I locked
my bedroom and turned to join him. For weeks he had refused to call me
anything other than '*Jjuko*,' my new Baganda title. The name identified
me instantly as a member of Tom's family clan, *Mborogoma*, the lions, one
of 52 totems in the tribe.

"It is a royal clan," he told me proudly. "From the same blood as the
Kabakas."

Passed on from a father to his children, clan bloodlines create a sense of
extended family within the larger tribal unit. Each totem, from mudfish to
colobus monkey, had legions of loyal members among the eight million
Bagandan people. Many of their traditional functions had broken down
since Obote abolished the *Kabaka's* court in 1966, but clans still played a
vital social role for the tribe, forming a basis for everything from sports

teams to arranged marriages and business deals. The cultural and political roles of Bagandan clans, however, stood to soon regain much of their former significance. In a move to bolster his popularity with several of the major tribes, Musevini had invited traditional kings to resume their thrones and ceremonial courts. Next week the Baganda would crown a new Kabaka, their first in nearly thirty years, ending the longest single lapse in a line of rulers stretching back more than six centuries, the oldest continuing monarchy on the African continent.

Tom and I met Dave Snedecker on the path, and we walked together through the wakening bustle of Kajansi on a Saturday morning. With no training sessions scheduled over the weekend, volunteers were free to relax, study, and explore the countryside around Kampala. Several times I joined the Ntales for their weekend activities, and today Tom had invited Dave and myself to help the *Mborogoma* clan prepare for next week's coronation ceremony.

Tom led us to a *duka* along the main road, and introduced us to the proprietress, a grandmotherly woman who smiled and brought out several stools from behind her counter. We thanked her and settled down in the shade to wait.

"The Ministry is sending a bus. They will pick us from here," Tom said, referring to his employers at the department of Transportation and Works. "They should come any time, depending on the petrol."

We sat in silence, watching cars and heavy old trucks throw back clouds of dust and greasy diesel smoke as they sped past. The Entebbe highway was a major thoroughfare, two lanes of blacktop crumbling away at the shoulders, a testing ground for the fleets of white Toyota vans that served as share-taxis throughout the country. Drivers raced and jockeyed along the busy route, honking and gesturing questioningly at anyone standing by the roadside. We saw dozens of people catch rides in both directions as the wait stretched into its second hour. I glanced at Dave and he rolled his eyes with a small shrug.

Living in Africa, I soon came to realize that impatience may be an entirely Western phenomena. Ugandans approach time from a fundamentally different perspective, not as a commodity to be spent or wasted, but as an event that unfolds continually, regardless of circumstances. Importance is placed on whether or not something happens at all, not on the timeliness of its occurrence. Tom showed no signs of frustration as the hours slipped by. We were going to the coronation site today, an event that was already in progress. Any amount of waiting incurred in getting there was simply a part of that event, part of the intrinsic process of happening.

Across the street, shop owners began opening for business, unlocking the metal grates in front of their stalls and sweeping debris out into the dusty roadside with brooms of bundled spear-grass. Street vendors unrolled mats directly onto the ground, and set their wares out before them, a varied collection of newspapers, old magazines, matches, chewing gum, pencils and sweets. Two men in stained white aprons struggled by with half a cow in a wheelbarrow. The freshly slaughtered meat glistened in the morning sunlight and left a trail of dark blood spots through the dirt. We heard the wet slap of a machete as they divided the haunches, then hung the best cuts on iron hooks dangling from the eaves of their shop. The ring of hammers on metal announced tinkers at work in the scrap yard, repairing bicycles, mending pots and shaping oil lamps from discarded cans. I watched one man weld an unfathomable series of rods onto a sheet of pig iron, squinting down at his work without any kind of a face shield, or even a pair of sunglasses. Some instinctive alarm in my mind linked welding flames with solar eclipses and *waragi* as a potential source of blindness, and I tried to look away. But like a moth to a light bulb my gaze crept back to that bright blue flare again and again as the hours dragged by, until the glance, the flame, and the turning away became my only markers for the passage of time.

Suddenly, Tom stood up and peered down the road. An ancient Mercedes bus careened into view and clattered to a halt before us in a plume of diesel smoke. Rust showed in patches through the cracked blue paint, but the words "Ministry of Transportation & Works" were clearly visible. Our ride had arrived.

Less than twenty miles separated us from our destination, but the trip took nearly two hours. We detoured through countless back-road trading centers, stopping to pick up dozens of Tom's co-workers from the Ministry. The occasion was obviously a major event, and many people were dressed in their traditional best, the women in bright satiny gowns called *gomas*, and the men decked out in long white smocks. Known as *kanjus*, the smocks had been worn for centuries, but were adapted in colonial times with the addition of navy blue blazers, giving men the aspect of students at some kind of Arabic/New England preparatory school.

Everyone joked and laughed as the bus rattled on, passing bottles of soda and a calabash of sour banana beer. The festive mood only faltered once, when the driver tried to collect money for fuel. People shouted him down angrily, and one man threatened to beat him. He skulked back to the front of the bus and drove on with the simple warning that he might not have enough gas to come pick us up.

"This man is not serious," Tom said in disgust. "The petrol was paid for by the Ministry. If it's not there, then he has sold it to his friends." As a landlocked nation, Uganda imports all its petroleum on tanker trucks from Mombasa, a Kenyan port on the Indian Ocean. Prices often approached four dollars per gallon, and the temptation to pilfer from government agencies fueled a small, but lucrative black market.

The usual throngs of pedestrian traffic increased to a solid mass of people as we approached the hilltop coronation site. The crowd parted slowly and the bus inched up the dusty track, passing under a series of bamboo archways decorated with bunches of banana leaves, red hibiscus flowers, and brilliant weavings of bougainvillea. The driver parked in a field near the crest of the hill, and we emerged into a mad sea of color and noise.

Thousands of *goma*-clad women rainbowed through the multitude and the air resounded with singing, cheers, and the staccato pulse of drumbeats. Tom led Dave and me directly to a line of hawkers and purchased three cloth *Mborogoma* badges.

"Every clan has a job here," he told us, pinning the labels to our shirts. "We will work with the lions."

Dave and I caused an immediate stir as we followed Tom through the crowd. Living in Kajansi was one thing, but *muzungus* attending a significant Bagandan cultural event produced a whole new level of amazement. People gaped and laughed, or rushed forward to shake our hands. Greeting them in their native tongue only exacerbated the situation, and our forward progress slowed to a near standstill. Tom introduced us as his American sons, and everyone thanked us profusely for coming to help them welcome their new Kabaka.

When they spied the roaring lions fixed on our chests, they would laugh anew, shouting "These are not *muzungus*, they are *Mborogoma*!" All day long, smiling people stopped me to ask, "What is your name? What is your clan?"

"*Jjuuko*," I would reply, and tell them my clan: "*Ndi Mborogoma*."

This response always produced howls of amusement, and someone would invariably buy a round of the closest beverage to hand. I began to wonder if Tom hadn't given me a clan name solely as a mechanism to obtain free drinks.

Dave and I worked our way slowly through the masses like a pair of comic celebrities, laughing along with the crowd and exchanging dazed looks. Finally, Tom led us into a large open space where construction was underway. At one end of the clearing stood a raised dais where the Kabaka would sit during the ceremony. Workers were erecting the poles of a thatched pavilion that would cover the platform and shade row upon row of chairs and benches for the government officials, clan leaders and foreign dignitaries expected to attend next week's coronation. An elaborate series of bamboo and grass fences, half-finished, surrounded and partitioned the entire site. Every structure was being built with traditional materials and designs, preserved in the tribe's collective memory for over fifty years since the last Kabaka, Frederick Mutesa II, received the trappings of office in 1939. The site already bore a strong resemblance to an early Bagandan court described in 1862 by John Hanning Speke, the first European to visit Uganda:

> *"The whole brow and sides of the hill on which we stood were covered with gigantic grass huts, thatched as neatly as so many heads dressed by a London barber, and fenced all around with the tall yellow reeds of the common Uganda tiger-grass; while within the inclosure the lines of huts were joined together, or partitioned off into courts with walls of the same grass."*

Famed for its elaborate hospitality and highly organized government, the Bagandan kingdom became a destination for early Europeans as the most 'civilized' kingdom in East and Central Africa, and eventually served as a center of British colonial influence. Explorers like Speke and Henry Morton Stanley counted Uganda as among the most hospitable places in the region, and spent weeks or even months as the Kabaka's favored guests (sometimes longer than they wanted, as the king was often loathe to let them go). After his first stay in 1875, Stanley wrote: "There is a singular fascination about this country. The land would be loved…even though it were a howling wilderness; but it owes a great deal of the power which it exercises over the imagination…that in it dwells a people peculiarly fascinating also."

The Baganda are a proud, even haughty race, and make up nearly half of Uganda's eighteen million people. Their homeland is the richest district in the country, and influential members of the tribe have often argued for an independent Bagandan state. President Musevini, a member of the neighboring Banyankole tribe, hopes that reinstating the Kabaka will appease Bagandan secessionists, but critics fear he is only adding fuel to the fire. Since colonial times, Ugandan leaders have struggled to balance the interests of the country against the demands of its largest tribe. Even

Idi Amin took steps to win Bagandan support. One of his first actions after toppling Milton Obote was to arrange proper burial for the remains of Mutesa II, who had died penniless in exile after Obote abolished his kingdom.

Although raised and educated in England, Mutesa's son, Ronald Mutebi, retains the devoted loyalty of his father's subjects. Excitement over his return had already reached a fevered pitch, and hawkers on the streets of Kampala did a brisk business in Kabaka posters, T-shirts and buttons. Mutebi, or 'King Ronnie,' as the press had already dubbed him, would be barred from political activity, but would surely exert considerable influence from his ceremonial throne.

"We are still Ugandans," Tom told me later, "but the Kabaka is our king."

Under the scorch and swelter of midday Ugandan sun, I began wishing the king had chosen a nice air conditioned palace for his coronation party. Tom scanned the grounds and led us over to where members of the Mborogoma clan were working on a long perimeter fence. Like many construction projects I witnessed in Uganda, building the Kabaka's fence involved seemingly countless numbers of people, and an incomprehensible amount of standing around. We joined a knot of the more active-looking workers, and helped bind bundles of bamboo and thick elephant grass to the support poles with strips of dried bark. But each flurry of activity led to an even longer period of loitering. With a few hard hats, orange vests, and a thermos of coffee, we could have passed for any city road crew in America. During one lull I peered down the rows of unfinished fencing that encircled the entire site, and realized that the *Mborogoma* clan would never finish their task in time for the ceremony.

It's not that there weren't plenty of workers available. More than 200,000 people milled through the coronation site that weekend, several thousand of whom were surely proud lion clan members. But most people had come only for the party—to dance, drink and socialize in celebration of their tribe's most important cultural event in over twenty years. They crowded around their long-neglected artifacts and holy sites: the huge, spreading canopy tree under whose branches the early Kabakas had first held court; a throne of living tree roots that recently reemerged from the earth in anticipation of the new king; and the gravelly hollow of a dry well that would only spring back to life after the King Ronnie received his crown.

"People will stay all week," Tom told me, but when I asked him if we would return for the ceremony, he said no. "From here you would see nothing. So many people. We will use the television." But television wasn't an option for most Baganda, and attendance at the coronation was expected at well over a million.

We made several trips across the compound, fetching bundles of dried elephant grass from the back of a large truck. But mostly we stood with the others, watching as our section of the fence slowly took shape. The sun beat down mercilessly, and I began envying the clans assigned to a job in the shade.

After some time, Tom made a show of wiping his hands, and gestured expansively around the hilltop. "Well, how do you see it?" he asked.

Following his gaze over the milling crowds and idle workers, Dave found the perfect summary: "There is very much work to do, I think." His command of Ugandan English and understatement was already approaching a masterful level.

"Mmm," Tom nodded in agreement. "Are you hungry? Let us find some lunch."

He set off and we followed him to the shady understory of a large jackfruit tree, where dozens of women crouched beside simmering cookpots of *matoke*, cabbage, beans, and ground-nut sauce. The women used pot lids and scraps of cardboard to fan the ruddy coals of their fires, transferring embers from one blaze to another with bare fingers, impervious to the heat. We crowded onto narrow benches with hundreds of other diners, who looked on, startled and amused to see *muzungus* eating with their hands. Someone passed around a banana leaf heaped with salt, and we dug in to our bowls of the hot, smoky stew.

After lunch and a cool soda, Tom led us back out into the crowds. We stood in line to catch a glimpse of the Kabaka's dry well and the throne of roots. We visited the ceremonial tree of justice, and everywhere we went people crowded around to thank us and make us welcome. Tom seemed to truly enjoy showing off Bagandan culture, and he reveled in the attention that Dave and I drew. We met bishops and politicians, and countless members of his clan and family as we meandered through the heat and crowds.

Later, Dave and I would remember that day as one of the most overwhelming of our time in Uganda, and one of the most revealing. The exuberant pride of the Baganda clans hinted at something fundamental in the fabric of their culture, a system of relationships woven like extended family throughout the populace. It lent a sense of intimacy to the masses of people, and I could think of nothing similar in America, where we describe a crowd as faceless.

"You should see the traditional dancing," Tom said, ushering us towards a throng of people singing and drumming in a tight group. Unseen hands pushed us to the front of the circle, where people took turns moving into the open, arms raised, shimmying their hips to the frantic drumbeat and chanting chorus. Several women wore fringed skirts over their *gomas* that rattled and waved with the rhythm, like loose sheaves of dry grass in a strong breeze. They danced with their heads thrown back, or looked down in concentration, but everyone was smiling, and the air felt charged and intimate. I watched, mesmerized by the sound and delirious pulse of the dance, until Tom pulled us away.

"You will learn that one at home," he assured me, and we moved back towards the slowly emerging Mborogoma fence.

All day long, a steady flow of people ascended and descended the hillside as if in pilgrimage, and the numbers at the top seemed undiminished late that afternoon when we returned to the roadside to find our homeward bus. There was, of course, no sign of vehicle or driver, so we moved into the shade of a nearby hedgerow. Someone produced benches, and we settled in for our second transportation wait of the day (This happens everywhere in Uganda. Rickety wooden benches always seem to be available at arm's reach, ready to appear whenever you feel the urge to sit down). Tom brought out a deck of playing cards and a lively game ensued as more of our fellow passengers appeared to wait for the missing bus.

To the northeast, windshields glinted with sunlight like a stream of tiny silver fish racing along the Entebbe highway, and Kampala spread its urban fingers across seven distant hilltops. People pointed towards the city, calling out their neighborhoods and arguing about the identity of various buildings. Which was Uganda Commercial Bank? Where was the Sheraton? Mulago Hospital? Mengo Palace? The Stadium?

Dave and I sat to one side, numb from the day's attention and relieved to be momentarily out of the limelight. On the slope below us I watched a flock of great blue turacos feeding raucously in the crown of a broad-limbed fruit tree. They looked unreal, like huge cerulean roosters, with drooping black cockscombs and crimson-tipped beaks. I watched them hop and jostle among the green branches, and tried to let time slip by unnoticed, to let sitting and watching birds be an active part of catching our transport back to Kajansi.

Maybe it worked, or maybe I dozed, but suddenly the sunset was hanging its lazy red banner over a distant hillside, and long shadows stretched across the coronation grounds. The bus had finally arrived, and as we drove slowly away, people were lighting huge bonfires to keep the festivities moving through the night. I saw the skeleton of our fence silhouetted in the firelight. No one was working, but the project was obviously well underway, and in that sense, right on schedule. The Mborogoma might not finish their task by next week, but eventually, the Kabaka would get his fence, just as surely as his rocky well would fill again with clear spring water, and his people would touch their foreheads to the ground at his passing. I had no doubt that these things would come to pass, all in good time. In African time.

CHAPTER IV

FREE SHOES FOR DOMINICO

▼

"Within seconds, the gorillas had breached the perimeter and trampled the mesh fence into the mud. Unchecked, they rushed into the compound, grunting and roaring. The driving rain matted their hair, giving them a sleek, menacing appearance in the red night lights. Elliot saw ten or fifteen animals inside the compound, trampling the tents and attacking the people..."

—Michael Crichton
Congo, 1980

The Peace Corps Uganda library contained only one book about gorillas, a battered paperback copy of Michael Crichton's *Congo*. The story took place in a remote rainforest north of the Virunga Volcanos, where killer albino gorillas stalked people through the jungle and smashed their heads in with stone ping-pong paddles. Realistic or not, this wasn't the most encouraging narrative for someone about to spend two years in a remote rainforest north of the Virunga Volcanos, looking for gorillas. But I read it anyway, and hung on every word.

For eight straight weeks, questions, thoughts and visions of The Impenetrable Forest had raced through my mind like a crazed mantra. Whether sitting through language lessons, building mud stoves, or haggling

over the next round of *munanasi*, some part of my thoughts wandered
through a forested hillside, chasing apes. All of the volunteers began chafing
at the restraints of life in Kajansi, where the Peace Corps tended to coddle us
like an overbearing parent. The last weeks of training grew fevered with
anticipation and speculation until we finally reached our five-day 'future site
visit,' a chance to experience first hand the various villages, schools, forest
reserves and national parks that we would call home for the next two years.

For some volunteers, the visit did little to bolster their confidence: "Go
to Iganga. Look for a Forest Officer named Joseph. He knows where your
site is, but you don't have a house yet." But most people glimpsed a life-
after-training that made us all the more eager to move up-country.

Bwindi Impenetrable Forest lay in the far southwest corner of the
country, a long and unreliable journey on public transportation. I had
only a few days to make the trip and was lucky to catch a ride with Liz
Macfie, my future supervisor from the International Gorilla Conservation
Program. She picked me up as early morning sun cast pale light over the
training center, and we set out west and south for the forest.

A British citizen raised in the States and trained as a veterinarian, Liz
was a self-assured, attractive woman in her mid-thirties. Before coming to
Uganda in 1992, she had spent three years tending to injured mountain
gorillas in Rwanda's Virunga Volcanos, staying on the job through periods
of unrest and outright civil war. Liz carried herself with the determination
of someone used to fighting for what she believed in. Ugandan wardens
and National Parks administrators secretly called her "the iron lady" for
her staunch defense of ecological and community conservation principles,
but she was respected and well-liked in most circles, with an easy style and
quick sense of humor.

"There's no dress code in Bwindi," Liz told me as I climbed into the
car, glancing sidelong at my Peace Corps regulation slacks and dress shirt.
"If you try to wear a tie, we'll cut it off." She then plugged the Rolling
Stones' "Sticky Fingers" album into the stereo, and I knew from that
moment that we were going to work together just fine.

Traffic thinned as we sped away from Kampala through a green land-scape of small *shambas*, swamps, and tangled woodlands. Papyrus reeds sprang up from the lowlands in dense, brindled fields, their tufted heads rippling in the breeze like plants from a Dr. Seuss story. Everywhere, brightly clad people walked along the roadside, leading livestock and car-rying bundles of goods from one town to the next.

The occasional vehicles that sped past us were heading for the city, heavily laden: semi-trailers, *matatu* vans, rusty tanker trucks, and private cars, all overloaded with people and baggage. In Uganda, the multitudes on the move greatly outnumber the vehicles available, and no space is wasted. Passengers crowded every seat and clung to the rails of pickups, lorries and flatbed trucks that were literally festooned with cargo—bunches of green *matoke* dangled from the fenders, crates and mattresses were stacked high on roof racks, and lake fish waved from side mirrors like huge satirical air fresheners. Every passing car looked like a compact gypsy caravan returning from the flea market.

"You really have to watch this spot," Liz said as we descended into another valley. "There's no sign to mark it, but the road is sinking into the swamp."

She slowed to thirty miles an hour, and drove carefully down the mid-dle of the road, but the whole car still rocked and shuddered as we crossed sections of pavement tilting sharply towards the muddy shoulder.

"How long's it been like this?" I asked.

"About a year, but it's getting worse."

We passed an overturned tanker truck, and a large area of blackened papyrus where something had gone over the edge and burned. I thought of Tom tallying paychecks down at the Ministry of Transportation, and wondered how many lawyers were on the payroll. Probably none. Uganda has Coca-cola, Rambo, Federal Express and a Sheraton, but the American passion for liability and lawsuits is a long way from catching on. People don't expect things to be fair, and unmarked road hazards are simply a fact of life.

White stripes on the blacktop marked the equator, and we continued south to Masaka, a city destroyed once by the war to oust Amin, and again when Musevini was battling to overthrow the Obote regime. Our road skirted the edge of the ruined town and took us west, further from Lake Victoria, where green faded quickly from the landscape. We passed through a dry rangeland of scrubby thorn trees and sun-scorched grass. Occasionally, a roadside market interrupted the miles of bleak earth and umber dust—tiny wooden stalls piled high with bright tomatoes, red onions, papaya, bananas and pineapple, as if the produce itself had stripped all color from the surrounding fields.

Twice, Liz had to brake hard and swerve to avoid cows on the road, the huge Ankole breed, with horns five feet from tip to tip. Skinny herdboys in ragged T-shirts and paper hats shouted and drove the cows away with long switches, still managing to stare in our direction, mouthing '*muzungu*' as we sped past. We stopped for lunch in Mbarara, a fast-growing city near Musevini's homeland, and drove on toward the far southwestern tip of the country. Near Kabale the road began climbing into the Kigezi highlands, a rugged terrain of steep, sculptured hillsides ringed with cultivation. Terraced fields contoured to every knoll like wrinkles, as if a stone had dropped on the top of each hill, and the hedgerows were tiny waves descending.

"Roll up your window," Liz told me. "We need carrots."

I smiled, and complied as if this made perfect sense.

Around the next curve, she slowed the car and pulled off at a small market stall. Suddenly, a crowd of shouting men and children surged up from the ditch, running towards us with armloads of fresh produce. They surrounded the vehicle instantly, waving vegetables and pleading with us to buy, buy, buy. The crush forced one man's face flat against my window, but he still managed to scream "Free for you," gesticulating with a large cabbage.

"Only carrots! *Karoti*!" we answered, and I inched the window open. Carrots were instantly squeezed through the crack, showering down onto my lap by the dozens.

"No! Just...want some...hey!" Hands forced the window down another notch and a sudden torrent of vegetables rained in: cabbages, cauliflower, squash, eggplant and sweet potatoes, all bouncing off the dashboard and rolling across the floor. Slapping hands away, I managed to close the window, while Liz sorted a few carrots out from the medley. We paid the proper salesman, and the others collected their unpurchased items with looks of sullen betrayal. Then another car pulled in behind us, and they dashed off in a mob, shouting and waving their wares anew. Only one vendor remained by the window, a forlorn-looking boy of seven or eight with two wispy bundles of spinach. "What about.... greens?"

Impulse shopping. We took a bundle of greens and hit the road.

In 1993, Kabale was a sleepy town. A civil war in neighboring Rwanda kept the border closed, and trade in the region had dwindled to nothing. Bicycles outnumbered cars fifty to one on the long, tree-shaded main street, and shuttered storefronts harkened back to a more prosperous time. We spent a night there before heading west again for the final, eighty mile trek to the forest.

The pavement ended half a mile outside of town, and we bounced onto a rough track of reddish dirt and gravel. The car rattled and shook, slamming over potholes and patches of bare rock like a carnival ride for masochists. After half an hour, it got noticeably worse.

"They've been working on the road," Liz explained with a shout. "This used to be really bad."

I could only nod as we clattered along and the road stretched out before us, a serpentine, red dust ribbon, climbing and descending the rugged landscape. Outside of the park itself, most native woodlands in the area had long since been cleared for agriculture, but stands of Eucalyptus dotted the hillsides, and we passed through the Mufuga Forest Reserve, a large plantation of Caribbean pines.

"We won't really see the park until we get to Buhoma," Liz told me apologetically. "The road circles all the way around the forest, just out of sight."

This was not a problem. After months of a lifestyle that combined the Peace Corps bureaucracy with a new understanding of the African sense of time, waiting had become habitual and a few more hours was child's play. And once my body adjusted to the road's constant kidney punches, I began to enjoy the verdant landscapes flashing past the window.

Cultivation crowded the valley floors, and small farms clung to the most precipitous hillsides—family compounds of thatch and tin-roofed huts surrounded by fields of banana, potatoes, sorghum and tea, brilliant green with new growth. The turned earth looked black and rich, a product of ancient volcanic activity in the area. The Kigezi, or Rukiga Highlands were formed by an upsurge of tectonic pressures along the western arm of the Great Rift Valley, where the restless Somali plate is slowly pulling East Africa away from the bulk of the continent. Twenty miles to the south, active volcanos in the Virunga chain still steam and gutter, spewing occasional eruptions of ash and lava into the surrounding countryside.

People waved and smiled from the villages and trading centers scattered along the route. With tourism still in its infancy, a car with two *muzungus* counted as a major spectacle for most rural farmers.

Two hours into the drive we crested a hilltop and saw the Great Rift itself dropping gradually downward into a wide, hazy expanse of grassy plains and woodlands. Blue hills in the distance framed the western edge of the valley, and to the north the waters of Lake Edward glinted like a band of fine silver.

"There's Zaire," Liz announced, and I stared across to a country with its own complex history of exploration and colonial folly. Under Belgium, the Congo Basin was ruled as a virtual fiefdom by King Leopold, who dreamt of building an economic power to rival Britain's East Indian Company. His Congo Free State was stymied by climate and logistics, a failure that helped cement Africa's reputation in nineteenth century Europe as 'The Dark Continent,' a sweltering wilderness plagued with fever and death.

Near Bwindi, the international boundary ran along the top of the rift escarpment, and most of the valley floor lay in Zaire, part of the huge but poorly protected Virungas National Park. From the Rift, Zaire stretched westward over the distant hills and a thousand miles of rainforest, winding down the muddy Congo River to the shores of the Atlantic Ocean.

Our route took us south now, near the border and up a steep-sided river valley. Cows grazed in the lowlands, a lush pasture sprinkled with *shambas* and shaded by occasional, massive canopy trees, remnants of the shrinking rainforest. Goatherds chased their animals from the roadside, and a group of men pushed bicycles laden with green *matoke*. We passed tiny villages whose names would soon become familiar, Rugando, Kanyashande, Mukono, and finally Buhoma, a cluster of huts where the road ended in a wall of treetrunks, tangled greens, and timeless shade. The Impenetrable Forest stretched away suddenly before us, towering and still in the midday heat, a serene curtain of emerald that filled the head of the valley and disappeared over the steep-backed hillsides.

I felt as if I'd stepped momentarily out of my body, replacing cognitive thought with a vapid smile and phrases like "Beautiful," "Wow," and "Looks great." In retrospect, the forest was magnificent, an intricate weaving of trees and vines that defied the eye to focus on any one aspect of the myriad whole. But at the time it was more overwhelming than the crowds at the Kabaka's coronation. Elation and awe vied with the suddenly imminent challenge of actually living there, of struggling to feel at home in a Central African jungle, and find a sense of place in a community of rural villagers.

I was grappling with this thought when Liz pointed up to a beautiful old church compound tucked into the hillside at the forest edge.

"There's your house," she said simply, and put the car in four wheel drive to climb the steep driveway. I felt like giving her a high-five.

The Apostles Church of Christ Jesus had transferred their flock to a neighboring village, and Liz used the abandoned church building as a combined residence and field office for the gorilla program. The pastor's house was mine, a three-room cottage with mud walls and a roof of woven banana fibers. Morning glory twined up the walls and into the damp thatching, framing the doorway in purple flowers. A narrow gorge separated our hill from the rainforest, and clear water ran through its fern-choked depths, filling the air with quiet stream-noise. Across the valley, Bwindi descended from the ridgetop in vast, looming waves of green.

We unloaded the car and Liz introduced me to Ephraim Akampurira, the gorilla program's houseboy, handyman, and occasional night guard. He would become a steadfast friend and my first point of contact with the village, but on that day he looked wary as he shook my hand with an almost imperceptible smile, "You are welcome."

My Peace Corps predecessor in Buhoma had gone home early, frustrated by the isolation of life in the jungle. She did, however, stay long enough to commission furniture and decorate the house. There were tables, chairs, a bookshelf, a bed, and two 'Greg Brady' style flybead curtains. She even left me a half-used bottle of imported Heinz ketchup; for a volunteer, that was the definitive sign that she had left in a hurry. I was unpacking my bags, setting books on the shelf and already beginning to feel at home when I heard a voice outside.

"*Kodi, kodiii,*" someone called. I walked into the yard and found Ephraim shaking hands with a short, cherubic old man in a frayed knit cap. "Hooo," he looked at me, chuckling as I approached. He grabbed my hand and beamed, chattering in Rukiga while Ephraim slowly translated.

"This one is your neighbor." He pointed towards a *shamba* some distance up the hillside.

"Dominico!" the man shouted merrily.

I told him my name, and he nodded with a knowing, thoughtful look, as if some secret plan was finally coming together.

"Ah, my son!" Dominico exclaimed in a sudden burst of English. "John...son. Liz...son. You? Son!" he cried again, still gripping my hand. I wondered for a moment if he was drunk, recalling a *muzee* at Annette's place who shook my hand for over forty minutes, chanting, "I love Clinton. **I loovve Clinn-tunn.**"

Ephraim was smiling now too, and laughing into his translation as Dominico went on to ask my age, my parents' names, my hometown, and whether or not I was married. He concluded with a solemn declaration that he would be my only father in Buhoma. Then he looked down with a sudden expression of sorrow.

"Eh, eh, eh," he made a kind of verbal 'tsking' sound, still staring at his feet.

"He says that he is suffering," Ephraim told me. "He says that when you go to Kampala you could bring him some shoes. Size seven."

Now we were all looking at Dominico's feet. They were gnarled, hardened and cracked from a lifetime of barefoot work in the fields. I felt suddenly conscious of my hiking boots and the pale flesh they hid, soft from a world where footware was taken for granted. Torn between compassion and the knowledge that I couldn't possibly shoe the whole village, I sought a compromise.

"Tell him I will look for sandals."

Dominico paused for a moment, as if considering the offer. Then he beamed anew, "My son! Good," and finally released my hand. He picked up his walking stick from the grass and shambled off down the hill, still laughing and shaking his head. I shook mine too, picturing the news of my arrival sweeping through town: "Free shoes! The Free Shoe Guy is here!"

At some point in training, the Peace Corps had warned us about this situation: "Don't give anything away. It sets a bad precedent." I felt like I'd failed an important early test, but realized later that in Dominico I'd come up against a master of negotiation. He struck early and aimed high, knowing full well that in a few weeks I'd argue for two hours over the stubby end of a pencil.

The next morning I woke suddenly with a deranged soprano opera blaring through the eaves of my hut. *Kuteerateera*, I thought in a panic.

Someone's being robbed. But as the screaming tapered off into soft hoots and pants, I realized that these cries were coming from a different set of neighbors: chimpanzees, announcing the start of another day in Bwindi forest. I lay there in the gray light before sunrise, listening to the chimps and the nocturnal chirr of insects. Gradually, night sounds gave way to a dawn chorus of birdsong: prinias and mannikins chattering in the undergrowth, the clamor of hornbills high in the canopy, and a red-chested cuckoo's descending three-note whistle.

I brushed my teeth and bathed outside as the sky brightened to saffron and rose, and slender fingers of mist drifted gently through the treetops across the valley. The chimps had moved further into the forest, but I watched a pair of cinnamon breasted bee-eaters preening in a tree fern, while a blue-headed agama lizard peered down at me from the roof-thatching, surreal yet pedestrian, like everything else in this strange new world.

At the park office I watched myself shake hands and exchange greetings with a legion of future co-workers and friends. *Betunga Williams, Phenny Gongo, Komunda, Kawermerwa, Yosamu, Rwahamuhanda, Mishana, Bafaki, Tibamanya, Tukamahabwa*...sesquipedalian names reeled past me in a spasm of garbled vowels, like music on a poorly tuned radio. I smiled a lot, and didn't try to pronounce anything. When two British tourists arrived with permits to track Mubare group, I found myself heading into the Impenetrable Forest for an urgent appointment: to meet my first group of mountain gorillas.

With one adult male, six females and five juveniles, Mubare was a stable, textbook gorilla family. The silverback, Ruhondeza, set the tone for the group. His name meant 'sleeps a lot,' after a local Rip Van Winkle story of a hunter who slept away three generations deep in the heart of the forest. The other gorillas followed Ruhondeza's calm lead, learning to tolerate daily visitors in less than two years. Group number two, the Katendegyere gorillas, were another story altogether. Bwindi's 'dysfunctional' apes included three silverbacks and a pair of feisty black-back males, all posturing and contending over two ladies and a child. Helping overcome their habit of screaming and charging to within inches of human observers was to be my first major task as a volunteer. But for day number one, my introduction to mountain gorilla behavior, I was more than happy to be visiting Mr. Sleepy.

We set out from the office in a flurry of walking sticks and parting handshakes, and I introduced myself to the tourists, a young British expatriate couple working for the World Bank in Kampala. Our guide, Medad Tumugabirwe, led us to the forest edge, under the spreading boughs of a fig tree, and launched into the park's standard tourist briefing. In spite of the shade and a cool morning breeze, he was sweating profusely, and I suddenly realized my presence was making him nervous. As a foreigner and his future supervisor, he assumed I must be a seasoned rainforest veteran, and an expert at tracking mountain gorillas. Not so. Beyond a basic degree in ecology, I had no real qualifications for the job, and sometimes wondered what determination the Peace Corps had used in my placement: "Well, he's a primate...he'll do."

But Medad didn't know any of this, and I kept a poker face, nodding sagely as he covered each point. It would be several weeks before the staff realized I couldn't track a gorilla in a bathtub, let alone the Impenetrable Forest.

"If the gorillas charge, you don't shout out or try to run," Medad concluded, looking furtively at the crib notes on his palm. "These gorillas are well habituated," he assured us, "but sometimes they can change their mind."

With that piece of information firmly in mind, we shouldered our gear and entered the forest. My first impression was shadow and wet, with pale treetrunks rising through the gloom like huge marble columns. Far above, a tight lattice of leaves and vines blocked all but the faintest rays of morning sun, and a cloud of pale butterflies winked like starlight through the shade. As my eyes adjusted, the dim undergrowth revealed itself in a chaos of greens—olive, emerald, and jade that permeated the air itself with a moist, earthy smell, like water dripping through moss. Dense thickets of saplings and tree ferns struggled upwards, vying for any patch of light, while creepers and lianas hung down from above in thick, ropy coils.

We soon veered off the main trail and the trackers disappeared instantly from sight. The forest definitely lived up to its names, both "Impenetrable," and "Bwindi," a 'dark, muddy place,' in the local tongue. We caught occasional glimpses of khaki raincoat through the leaves, but mainly followed the trackers by the metallic ring of their *pangas* as they cut a narrow path through the vegetation. Several hours passed with slowly building anticipation as we climbed and descended, retracing the gorillas' meandering route across the steep hillside. Medad pointed out trampled leaves and broken saplings along the way, assuring us that, "Mubare group has passed here, but yesterday."

Once, the canopy above us erupted with shaking branches and high-pitched, bird-like chirps. A flash of russet fur and the bearded face of a red-tail monkey peered down for an instant before leaping away into the green. We pressed on and came finally to a cluster of carefully matted leaf-and-vine platforms, set low in the vegetation of a small clearing.

Mountain gorillas construct a new bedroom every night of their adult lives. At dusk each individual pulls down the surrounding branches and leaves to weave a stable, circular nest from four to five feet in diameter. Usually grouped loosely around the lead silverback, these night nests help keep the apes from sliding downhill while they sleep. On the rare occasion when a gorilla decides to sleep in the trees, nesting prevents its massive bulk from tumbling earthward during the night. Only infants and juveniles escape this daily task, choosing to sleep with their mothers until fully weaned after three to four years.

As we approached, the trackers were bent over, examining the nests carefully. Something was wrong.

"No dung," Medad whispered. "They did not spend the night."

Gorillas show far more aptitude for construction than housekeeping, and invariably soil their beds at some point during the night. Measuring dung width and counting nests is an important technique for censusing wild populations, but scientists would have found little to go on here. The nests were perfectly arranged, but there was no sign of the distinctive three-lobed bolus. Mubare group had settled in for the night, but moved on before sleeping. Something had driven them off.

"*Empazi*," George whispered urgently, pointing down among the leaves. We followed his gaze and suddenly saw the ground seething. A thick column of safari ants swarmed across the clearing, millions of red-bodied workers flanked by lines of huge 'bulldogs,' equipped with long, menacing pincers.

"Ants!" Medad translated needlessly. "Hurry!"

We high-stepped across the glade, stamping our boots and brushing ants from our legs in what was probably a perfect re-enactment of Mubare group's exodus the night before. Cursing and slapping, we escaped with only a few needle-prick bites, but something about the practiced way Medad and the trackers plucked ants from their socks told me this wouldn't be my last encounter with safari ants in Bwindi. We pressed on and soon came to another set of nests, complete with requisite dung, a few hundred yards up the slope.

From there the trail grew rapidly fresher, with redolent, barnyard-smelling spoor and leafy stems still wet from chewing. The trackers advanced carefully now, their *panga* swings barely audible as we snuck up the hillside. Suddenly, a new sound brought everyone up short: a single grunt, deep and clear, the way a malamute might bark if it weighed four hundred pounds.

"*Engagi?*" I asked, whispering one word of the local dialect I'd been sure to memorize in advance.

Medad smiled and nodded, "Yes. A gorilla."

We continued slowly up the slope and the trackers started making low, coughing noises, an imitation of gorillas feeding or at rest. I glanced back at the British couple. They looked sweaty and mud-streaked, but wired with excitement, eyes as wide and expectant as my own felt, like cartoon aviator goggles stretched across my face. Another grunt and a long sigh sounded from up the hill, but closer, much closer.

"They are moving," Medad told us. "We must be careful."

Suddenly, a shadow detached itself from the greater dimness, a brief image of ponderous grace glimpsed through gaps in the foliage. My first thought was that it must be the lead tracker, so human were the movements. But then she came fully into view, walking hunched forward on her knuckles with a baby clinging to the black ridge of her back, and all thought fled before awe and a far less profound realization: *HUGE. REALLY HUGE.*

She strutted past and out of sight without a glance in our direction, but the baby's head turned, regarding us with a long, brown-eyed blink before disappearing into the green. Then the undergrowth rustled and a pair of juveniles tumbled down the slope, two stout, round bodies covered in thick hair. They pawed at each other with teeth bared in mock smiles, rolling through the leaves like shaggy inverse snowballs. Their play brought them eventually to a dark mound in the undergrowth, and they climbed up to begin a series of daredevil leaps into the surrounding branches. When one of the youngsters began to wander in our direction, the dark mound reached out and gently pulled it back.

"Silverback," Medad hissed, and the shadow coalesced before my eyes into the prostrate form of an even larger gorilla, twenty feet away through the bracken.

Ruhondeza lay sideways with his head facing towards us and feet stretched out of sight behind the wide bulge of his stomach. I watched his children clamber back across one burly arm, and had to recalibrate my entire notion of gorilla-scale: torso like a tractor tire, tree-trunk legs, and head like a scrap-iron barrel, carved with a Rushmore face. Adult males average twice the size of their female counterparts, and even lying down, Ruhondeza carried every pound with a looming sense of authority and serene grace.

We watched for several minutes as the juveniles wrestled and played around him, a tranquil family picture interrupted only by the sound of Ruhondeza's occasional, rumbling sighs. Huddled together and hemmed in by walls of vegetation, the scene felt intimate and infinitely tender.

I stayed in the back of the group, watching through a narrow gap in the foliage while the tourists crowded forward, snapping photo after photo, and struggling to hold their cameras steady in the weak light. Their shutter noise eventually disturbed Ruhondeza's slumber and he shifted, turning his massive, peaked head to look in our direction. I met his gaze through the undergrowth and we both moved sideways to get a clear view. Finally he looked away, and I sat back in wonder, disbelieving that this would all become a part of my daily routine, that soon I might see recognition as well as curiosity in those calm, intelligent eyes.

CHAPTER V

DANCES LIKE A CHICKEN

▼

*"Love carefully. The person you so cherish
might be the special one to lead you to your grave."*

—Uganda AIDS Commission Billboard, *1993*

"The man with the radio who dances like a chicken. He is quite mad, I think."

"Yes, and a drunkard!" Tom bellowed, laughing as the bent, skinny figure lurched across the dirt courtyard, clutching a shortwave radio and strutting in time to its noisy blasts of static.

Late morning sun glared down from a cloudless sky, casting sharp-angled shadows across the benches outside Annette's place. We sipped orange sodas, waiting to meet a group of Tom's neighbors and share a taxi ride to the village home of Susan's parents. It was my last full weekend with the Ntales before moving permanently to the Impenetrable Forest, but our activities promised to be far less festive than royal fence-building. Susan's younger brother had died the previous morning, and we were on our way to the burial.

"He was only twenty years, but very sick," Tom had explained. I thought immediately of AIDS, but said nothing. In Uganda, where researchers estimate between ten and twenty percent of the population is

HIV-positive, everyone has lost friends and relatives to the disease. Posters and billboards throughout the country preach safe sex, and people talk constantly about the hope for a cure, but rarely in reference to any specific person. When it strikes close to home, the virus is still an uncomfortable topic.

"Susan will be gone for some days," he went on. "To be with the mother."

Traditionally, neighbors and friends stayed for a continuous three-day vigil at the family home, so the immediate relatives would never be alone with their dead. The men built bonfires every night to drive away spirits, while the women gathered inside, cooking to feed the guests, and preparing the body for burial. In recent decades, the sheer number of deaths has necessarily led to a shorter mourning period. Closer friends and relations will still stay with the family, but most return home after a single night.

During my time at the Ntales, Tom and Susan attended funerals, burials and last rite ceremonies at least once a week. "We have lost a neighbor," Tom would say, or, "It is a friend from the church." Twenty years of oppression and civil war, followed by the worst AIDS epidemic on the continent, has made death a routine part of daily life in modern Uganda.

"That's how it is to live here," Tom told me one morning, bleary-eyed from another all-night vigil. Attending funerals is a social duty, but people regard most deaths practically, with phlegmatic acceptance, reserving true grief for only their closest family members and friends. Their stoicism arose from necessity, but its history extends far beyond HIV and the Obote/Amin years.

For untold generations, Ugandans lived a lifestyle of subsistence hunting and farming, plagued by malaria, tribal warfare, and the sometimes-tyrannical rule of their traditional kings. While admired for establishing complex court systems and organized administrations, Uganda's local monarchs often governed with brutal authority and little regard for the lower classes. During his stay in 1862, Speke witnessed numerous executions in the court of Kabaka Mutesa I. The king showed a great interest in the killing potential of European guns, ordering the death of his subjects as casually as he shot the cows and birds roaming the palace grounds:

> *"The king now loaded one of the carbines I had given him with his own hands, and giving it full-cock to a page, told him to go out and shoot a man in the outer court; which was no sooner accomplished than the little urchin returned to announce his success, with a look of glee such as one would see in the face of a boy who had robbed a bird's nest, caught a trout, or done any other boyish trick...I never heard, and there appeared no curiosity to know, what individual human being the urchin had deprived of life."*

Sir Gerald Portal, leader of Britain's mission to Uganda in 1893, noted that "In the days of the late King Mutesa, and during the first years of the reign of Mwanga, executions on the most trivial pretext were of daily occurrence, not only at the capital and by command of the king, but all over the country, and at the mere will of these *batongoli*, or district chiefs." Common people didn't fare much better in the neighboring kingdom of Ankole where the *Bahima*, or ruling class, were said to never touch the ground, and would ceremonially walk across the backs of prostrate peasants and spit into their open mouths.

Coming from this history of subjugation, Ugandans have viewed more recent trials like Colonialism and Idi Amin as simply the latest in a long series of hardships. In the case of AIDS, people are aware of the threat, but they don't live in fear. A person could struggle to avoid HIV, but still die from malaria or tuberculosis, or there could be another war. Aunt Florence summarized the viewpoint perfectly one evening over a cup of wine. She had just received news of a death in the neighborhood, and shook her head with a sad laugh. "Life is short," she told me, raising her glass, "so drink up."

The NRM government has attacked the AIDS crisis with a straightforward public education campaign. Billboards in every town and posters throughout the public schools may be having an effect on the younger generation, but adults have been slow to change their habits, and the epidemic continues to grow through long-established customs of polygamy and widespread extramarital sex. I once asked Tom how many years he and Susan had been married, and he looked shocked.

"But we are not married," he exclaimed, amazed I could have misinterpreted the situation. He went on to explain how he had eight children from seven different women: "My first job with the Ministry was in Masaka. For two years I stayed there, and I produced a kid. Then I was transferred to Kyambogo-side, this small place going towards Jinja. There I produced another kid. Then I shifted to Kampala…" His progeny were scattered throughout south-central Uganda and ranged in age from two to twenty, but he assured me that he had never been 'officially' married. "You have one legal wife whom you marry, but these others," he made a dismissive gesture, "what are they?"

We finished our sodas and Tom bought a round of *munanasi* before his friends finally arrived. By the time we all reached the funeral, more than 200 people were already crowded into the small, dusty yard. Susan's family home stood close to Entebbe, a stout farmhouse with rusted tin roofing and mud-brick walls, set back from the highway among tall mango trees, narrow fields of arrow-root, and a stand of drooping bananas. The mourners crouched on straw mats and sat on benches in any available scrap of shade, talking softly while a hot breeze rifled the mango leaves above, and cars rushed by on the busy Entebbe Road.

We joined several men in the wavering shadow of a ragged banana tree. Tom's bearded face was fisted and turned down in a stern frown, and he dabbed sweat from his forehead with a thin handkerchief. The rest of the group grew somber as well, from sadness perhaps, but also because we were too late to find good seats in the shade.

Across the yard I spotted D.K., a regular at Annette's place. He was standing near the house with a cut banana leaf balanced on his head like a long, green-billed hat. When he saw me he smiled with a friendly wave, the grin incongruous, like something neon among the sea of dark, solemn faces.

Before the service, Susan emerged from the house and came over to greet us. She knelt carefully, keeping the crimson folds of her *goma* out of the dirt. We wouldn't meet the rest of the family, Tom explained, because it went against custom to socialize with your in-laws. Baganda tradition forbids men from even stepping under the same roof as their spouse's mother, an ironic twist on a long-running joke in Western cultures.

Susan returned to the house as they brought the simple wooden coffin outside, setting it carefully in the middle of the yard. A black-shirted priest followed, and everyone bowed their heads for the first prayer. Two hens and a sickly rooster pecked the ground at our feet, clucking and squabbling until someone shooed them into the cookhouse and shut the door.

The service was Catholic and long, a sermon and bible readings in *Luganda*, interspersed with prayers and singing. There were no hymnals. People knew these verses from long practice, and seemed to anticipate each song before the pastor could even announce it. They sang loudly, but without enthusiasm, staring unfocussed into the distance. I stole glances at Tom, but he too looked disinterested, even bored as the funeral dragged on. Everyone prayed and chanted mechanically until the whole ceremony had the feeling of mourning by rote. As the coffin was lowered into a grave behind the family home, the mother and two sisters suddenly lunged forward, wailing and keening, but without tears.

I watched with curious sympathy, struck by memories of another funeral I'd attended recently, for a childhood friend who died in a house fire shortly before I left the States. At his service, the shock and grief were palpable. Our sorrow was tinged with outrage, pervaded by the feeling that death had come too early, and cheated him out of the best years of his life. In Uganda, however, people don't expect to lead long lives. Where the average person reaches only 42 years of age, any man in his mid-thirties could expect to be called *muzee*, 'old sir,' a title of honor. Elderly people received even greater respect, in part from a simple appreciation that they had managed to survive their lives.

With the body laid to rest, the pastor asked for donations to help the family, and everyone dropped something into his basket before filing out of the yard. Tom and I joined D.K., Vincent, and most of the regulars to find transport back to Kajansi. The men began laughing and carrying on in high spirits as soon as the funeral was out of sight behind us, and we went directly from the taxi park to a bench outside Annette's place.

The skinny drunkard was still there, jerking around an alleyway with the unflagging stamina of true madness. In Uganda, each town had its cadre of the visibly unstable. With little professional treatment available, they simply became a part of the village landscape, cared for by extended family and the community in general with an almost affectionate acceptance. Some even gained a certain regional notoriety, like Mbarara's man in burlap, the stone-throwing woman in the Ishaka taxi park, or the eternal hitchiker, doggedly flagging every passing vehicle from his roadside village near Masaka. Nobody bothered the dancing man and his radio. In time, someone would probably bring him a meal.

Annette greeted us with a jug of *munanasi*, and we launched into a boisterous round of *matatu*, the local card game. It's a kind of complex crazy eights played at lightning speeds, and named for the equally breakneck taxi vans that ply the Entebbe road. As usual, we divided into clan-based teams, Tom and I cheering "*Mborogoma!*" whenever we won a hand.

Coming immediately in the wake of a funeral, the scene's brazen jollity felt contrived, almost eery. Nobody mentioned the deceased, or talked about the service. If there was any lingering melancholy, I seemed to be the only one feeling it, and I hadn't even known the man. But he had been close to my own age, and I found myself wondering about him, wondering who was mourning his loss. Finally, my curiosity got the best of tact, and I asked Tom directly if Susan's brother had died of AIDS.

His face went stark and solemn in an instant, as out of place in the bustle of the bar as D.K.'s smile had been at the funeral. "Yes," he said and finished his wine in a gulp. "This one is terrible." He paused for a moment, then turned towards me, taking in the whole bar with a nod of his head, "Look around you."

Michael and Vincent were arguing happily over the cards. A young boy walked by, selling sodas and skewers of roasted meat. Annette's baby sat in the dirt, playing with a broken sandal, and the chicken man had finally stopped dancing; he stood stock-still, with the radio pressed against his ear like a shrieking plastic seashell.

"Everyone you see here is carrying the virus," Tom said quietly, "Everyone." And in his suddenly haunted eyes, I glimpsed for a heartbeat all the poignancy and grief that seemed so absent from the funeral. We held that gaze for less than a second.

"I will have you, *Mborogoma!*" D.K. laughed, calling us back into the game.

Tom blinked and his smile returned in a flash. "*Mborogoma!*" he shouted back, and poured wine into our empty cups while D.K. shuffled, called the deuces wild, and dealt a new hand.

CHAPTER VI

SORRY, SORRY SORRY

▼

"This roar was the most singular and awful noise I had ever heard in these African forests. It began with a sharp 'bark,' like that of an angry dog; then glided into a deep bass roll which literally and closely resembled the roll of distant thunder along the sky. I have heard the lion roar, but greater, deeper and more fearful is the roar of the gorilla... The earth was literally shaking under my feet as he roared, and for a while I knew not where I was. Was it an apparition from the infernal regions? Was I asleep or not? I was soon reminded that it was not a dream."

—Paul du Chaillu
Wildlife Under the Equator, 1869

My first major task in Bwindi Forest was to habituate the Katendegyere gorillas. Mubare group had been receiving paying visitors for several months, and that revenue awakened park headquarters to the potential of gorilla tourism. Pressure from senior officials and increasing demand for tracking permits led my marching orders: have the stubborn, aggressive apes of Katendegyere group calm enough for tourists by the end of the year.

As I walked towards the office one morning through birdsong and early grey light, I found myself hoping, as I hoped every day, that the trackers would be hungover.

After years of chasing animals through the forest, Prunari Rukundema, Charles Kyomukama and Mishana James no longer differentiated between level ground and Bwindi's precipitous, mud-slick hillsides. They set a grueling pace up every incline and skipped nimbly down the slopes, wearing simple flip-flop sandals or plastic boots, with lunch pails balanced easily on their heads. I felt ridiculous flailing along behind them in my hundred-dollar hiking shoes and fancy rain gear, slipping and tumbling over tree roots and shrubs like a giant Goretex pinball. Only Charles spoke any English, but they all soon learned to express their sympathy as I crashed and slid through the undergrowth.

"Sorry, sorry, sorry!" came their cries after every fall, a distinctive chorus that marked our path through the jungle as surely as raucous croaks announced the hornbill, or a screech the panicked monkey.

My only chance to keep up lay in their fondness for *tonto*, the local banana beer. On mornings after a paycheck, market day, or any other time they could afford to tie one on, their speed through the forest slowed from breakneck to merely brisk. Sometimes, if *waragi* had been involved, they were still staggering, and I was able to get in a few 'sorrys' of my own.

That morning, I knew I was in luck when Prunari greeted me with bloodshot eyes and a mumbled, boozy-smelling "*Agandi.*" He and Charles set off at a weary march while I brought up the rear with Mishana, a tall, impossibly skinny man whose face knew only two expressions: pinched in nervous thought, or leering a yellow-toothed grin. Today he was thinking as we started the long, vertical climb up Rushura hill.

After several quiet minutes, we heard a shout and Phenny Gongo caught up with us on the trail. Smiling and sober, he taunted the trackers in Rukiga before moving up to take the lead. With a strong work ethic and an excellent command of English, Phenny was one of Bwindi's best ranger-guides. His mischievous sense of humor transcended culture and endeared him to the staff and tourists alike. We'd become friends easily, and he often used my Peace Corps address for the postcards, letters and packages his new international acquaintances began sending him from around the world.

A month before my arrival, Liz received a formal visit from Phenny and Hope Nsiime, our only female guide. After tea and polite, roundabout greetings, Hope and Phenny turned to Liz with a curious hypothetical question: would the park or the gorilla program see any problem with having two ranger-guides who were married? Liz happily answered no, and the wedding went on to become one of Buhoma's most legendary parties. Hope and Phenny were now expecting their first child, and talked eagerly about opening a restaurant across from the new campground.

"The trackers are weak," Phenny called back to me, laughing, and swinging at an overhanging branch with his *panga*. "They are paying for last night!"

"Ah, Gongo," I answered, calling on one of my newest Ugandan-English phrases: "On Rushura, we are all suffering together."

At more than 6,200 feet, Rushura Hill loomed high over Bwindi's western park boundary and the Zairian border. The trail ran straight up through steep-banked banana *shambas* and millet fields along the park edge. With a hot tropical sun rising overhead, climbing without the benefit of shade seemed like suffering indeed, a penance that Katendegyere group put us through nearly every morning. They had lingered for close to a month in a shrub-choked valley, two ridges south of the summit. Eventually, my legs grew used to the climb, but in those first weeks I gained a lasting empathy for every tourist who followed in the years ahead; some of them would turn around and give up their tracking permits after one look at the towering hillside.

On the final incline we passed a row of tall Caribbean pines, an introduced species planted as boundary markers in the 1960s, when Bwindi was still a Forest Reserve. The smell of pitch, and the hiss of wind through needles stood out from the humid green of the rainforest like glaring stage-props. But at times their incongruity was comforting, an evergreen reminder of my home in the Pacific Northwest. As if to complete the picture, slaty dark rainclouds moved in to cover the sky when we reached the summit.

We paused for water near the stump of a small tree, where someone had crossed into the park cutting firewood. Only three years earlier, fresh stumps were all too common in Bwindi. Corruption in the Forest Department led to unchecked poaching, and made illegal logging one of the area's top industries. The struggle to upgrade Bwindi from Forest Reserve to National Park status came at a critical time. Large mahoganies and other valuable hardwoods had been extracted from nearly 90% of the forest, while poachers had hunted leopards and buffalo to local extinction, and reduced the elephant population to fewer than twenty-five individuals. Lackadaisical rangers allowed people to encroach and settle inside the Reserve boundaries, and long term prospects for the forest looked bleak.

Gorillas had survived in Bwindi for one primary reason: local people do not regard them as food. Unlike in West Africa, where apes are considered a delicacy, and prized for their medicinal uses, Ugandan hunters tended to fear and avoid the large primates. In the late 1980s, nearly 300 gorillas remained, living in obscurity in the rugged, high-altitude portions of the park. Although separated from the well-studied Virunga population by only twenty-five miles, Bwindi's gorillas were virtually unknown, and their taxonomic status had never been established. When a genetic study showed them to be mountain gorillas, rarest of all the great apes, conservationists had at last found their rally cry.

By focusing on the gorillas, an endangered and famous species, Ugandan and International groups helped urge the government to protect all of the Impenetrable Forest, and its unusually varied collection of flora and fauna. As one of the few rainforests in Africa to encompass both lowland and montane habitats, Bwindi gives home to an exceptional array of wildlife. Preliminary surveys indicate more than 350 bird species, between 300 and 400 types of butterfly, over 220 trees, and 100 different mammals, including ten primates. For the sake of comparison, Bwindi's biodiversity surpasses that of all six New England states combined, yet is found in a single forest less than a tenth the size of the greater Boston metropolitan area.

When the Musevini government gazetted Bwindi, the Rwenzori Mountains, and nearby Mgahinga Gorilla Reserve as national parks in 1991, they demonstrated an encouraging commitment to protect Uganda's natural resources. The move earned praise from the international community, and saved vital habitat for gorillas and myriad other species, but not without certain expectations. Musevini had pledged to revitalize Uganda's tourism industry, and certainly recognized the economic potential of gorilla tracking, which, in times of peace, had been the third highest generator of foreign exchange in neighboring Rwanda.

Many in the scientific community had hoped to preserve Bwindi's mountain gorillas in a completely undisturbed setting, an exclusive preserve for a fragile, endangered species. The same arguments surfaced in the early 1980s, when legendary researcher Dian Fossey opposed the first gorilla tourism efforts in Rwanda and Zaire. But while the concept of a 'gorilla ark' is appealing to nearly all conservationists, the idea rates as impractical, even untenable, in modern Central Africa. Both Bwindi Forest and the Virunga Volcanoes are tiny islands of habitat, surrounded by one of the densest human populations on the continent. The area already supports between 100 and 450 people per square kilometer, and those numbers are expected to double in the next twenty years.

Faced with this inevitable, increasing demand for land and forest products, local people and governments need very tangible reasons to set aside parks and reserves. Natural resource managers must embrace every possible conservation tool, and Uganda National Parks decided to center its efforts in Bwindi around the economic benefits of a carefully managed tourism program. In 1992, they appealed to The International Gorilla Conservation Program (IGCP) for advice. Sponsored jointly by The World Wildlife Fund, The African Wildlife Foundation, and Britain's Flora and Fauna Preservation Society, IGCP brought financial support, as well as mountain gorilla expertise in the form of my Buhoma colleague, Dr. Liz Macfie.

Drawing on her experiences as a gorilla veterinarian in Rwanda, and the results of gorilla tourism in Zaire, Liz helped design a project in Bwindi that would emulate those countries' financial successes, while avoiding the pitfalls of over-habituation, excessive behavioral disturbance, and the risk of human disease transmission.

"In Rwanda, we had generations of gorillas growing up in close contact with people," she told me, referring to the apes habituated by Dian Fossey in her pioneering 1970s study. "People didn't know about disease transmission then, and they regularly came into physical contact with the gorillas. Now those animals are set in their habits, and their health is at serious risk; not to mention the danger to people. When a forty-pound juvenile crawls onto your lap, it's not a problem. But when that gorilla grows up to weigh 350 pounds, his play behavior can break your arm."

The challenge lies in finding a balance, getting close enough to observe the gorillas without exposing them to human diseases, or having too strong an impact on their natural behavior. Habituated animals should ignore people as much as possible, treat them as an innocuous part of the landscape. When they begin interacting with their human observers, the process has gone too far and puts both parties at risk. With Katendegyere group we were still in the early 'patience' stages, trying to find a distance at which the gorillas were comfortable, then gradually closing that gap until we could view them regularly through Bwindi's dense undergrowth.

While I was habituating Katendegyere, Liz and the park advisory committee were hard at work editing Bwindi's first management plan. But strict safeguards for the gorillas had already been put in place. We maintained a minimum distance of five meters from the animals at all times, and refrained from eating, drinking or talking above a whisper within a quarter-mile. All trash and food scraps were packed out, and any human waste buried at least ten inches underground. Tourists visiting Mubare group were limited to six people in a tight, quiet group, accompanied by a park guide and trackers. They stayed with the gorillas for a maximum of one hour, and anyone exhibiting signs of illness would be turned back, including park staff.

If there was any statistical risk of gorillas catching a human hangover, the trackers would certainly have stayed behind this morning. But as the hike stretched into its second hour, they began to recover, bantering back and forth and slowly picking up the pace. We entered Bwindi's dim shade near the hilltop, and moved quickly now, grabbing at branches and tree roots for support as we descended a steep, slippery trail to the valley floor. A shallow gulch trickling with water marked the source of Muzabajiro Creek, and we crossed it in easy leaps. Soon we were climbing again through a large clearing in the forest, where isolated canopy trees lunged skyward like spreading hands, surrounded by thick waves of dense green shrubbery. We found Katendegyere group's day-old trail near the top of the second rise, and turned to follow, cutting back the vegetation with ringing *panga* blows. Our passing sent a troop of blue monkeys leaping through the treetops, and startled up clouds of tiny indigo butterflies that winked through the leafy green around us like specks of misplaced midnight.

When the trail began to look fresh, Prunari and James took the lead. They were both talented trackers, able to follow the apes through any terrain, but cursed by one of the cruelest career-choice ironies I've ever encountered: they were terrified of gorillas. To be fair, Katendegyere group's five males often struggled against one another for dominance, and were notoriously nervous around people. I hadn't seen more than a few lunges and barks, but I knew their aggression could easily lead to a charge on their human observers.

Still, the group's habituation might have progressed faster without the trackers' habit of avoidance. Without supervision, they were known to hike into the forest, have lunch and a quick nap, then return to the office with a glowing report that "the gorillas are very fine." My major role on the team was simply to ensure that we reached the group, and actually stayed with them for the full habituation period.

I wanted the process to be successful, but at times I felt terrible putting the trackers through it. Their anxiety increased visibly the closer we came, and I tried to imagine their thoughts as they deciphered the complex bent-leaf trail: "*Ah, see the broken twigs there, the gorillas have passed this way. Here is the dung of a male. The trail is becoming very fresh now. SHIT!!!*"

We descended slowly into a narrow, knife-shaped ravine and soon enough heard the twig-snapping racket of feeding apes. Just then, a celestial valve opened somewhere high in the clouds, and rain began tumbling down over us in an unbroken lead glass curtain. We glimpsed flashes of black as the gorillas moved deeper into the undergrowth to wait out the storm, and we cast about to find our own shelter from the rain.

As we pulled up our hoods and hunkered down, Phenny touched my arm and gestured across the ravine. A gorilla, one of the females, had settled in full view less than thirty yards away. Through binoculars I could see the water running down her face as she stared resolutely into the storm. Her eyes held a look of calm wisdom, like a patient old fisherman, or a veteran schoolteacher—someone in control, with little room for surprise.

I pulled out my notebook and tried sketching her 'nose-print,' the distinctive pattern of wrinkles and scars that researchers use to distinguish individuals in a group. Recognizing specific animals is essential to understanding social dynamics and predicting an individual's behavior. With the trackers' help, I hoped to identify and name all the Katendegyere and Mubare gorillas, and Liz had asked me to establish a set of photographs for the IGCP files. But with rainwater sluicing over the lens of my binoculars, I gave up on art and photography, and settled in to wait, unconsciously mimicking the sullen posture of my primate cousin across the valley.

The cloudburst came down in heavy, tropical drops that seemed not so much to fall as to bounce, setting every individual leaf into spastic motion, as if the whole forest was orchestrated in a kind of grand, botanical dance. After an hour, the shower finally began tapering off. We stood up to shake the water from our raincoats and watched Katendegyere group slowly rouse themselves to action. But in the ravine's thick tangle, we didn't see gorillas so much as the movement of gorillas, a shaking branch, shuddering leaves, or the tall shrubs parting from below as individuals spread out around us to feed.

The female had moved closer, and squatted now on a wide log, stripping thin creepers from an overhanging branch. As she reached up to pull down a fistful of vines, something drew my attention to her right hand. The last two fingers stood out at a sharp angle, broken some time ago and improperly healed. The injury may have resulted from a fall or a fight, but more probably arose from one of the many wire snares set in the forest by poachers hunting bushpigs, duiker,[*] and other small game. Wound tight around a gorilla's foot or wrist, snares often lead to serious infections, gangrene, and even death. With only two fingers maimed, this gorilla had gotten off lightly. They actually made her appear almost quaint, handling each leaf daintily, like an English matron serving tea in delicate porcelain cups.

I turned to ask the trackers if she had a name, but before I could speak, the foliage behind them erupted, and a large male gorilla charged full speed down the slope. He moved in a hunched, loping gallop that should have seemed awkward, but managed to cover the ground between us with lightning fluidity. Screaming in rage and trampling everything in his path, he was on us in something less than a heartbeat.

Renowned zoologist George Schaller once wrote that gorillas "are eminently gentle and amiable creatures, and the dictum of peaceful coexistence is their way of life." *What an idiot*, I thought. Somewhere in my mind, I knew that charging was only a bluff tactic, an elaborate display that actually helped minimize direct physical conflict between males. All the textbooks agreed that so long as I didn't fight back or try to run, the actual danger was negligible. But with a roaring, furious

---

[*]   Bwindi harbors at least three species of duiker, the diminutive antelope common in Central African forests. Adults range from rabbit-sized to more than 175 pounds, depending on the species.

gorilla thundering towards me, this knowledge fled from my brain faster than goats from a sleeping herdboy.

The ape careened past in a flash of bared teeth and wild eyes, less than an arm's reach from my face, and in that instant I gained complete understanding for the trackers' fear. My rational mind cowered behind one overpowering impulse: *let's get the hell out of here.*

The gorilla spun away, still screaming, and continued down the slope. The noise alone was heart-stopping: an indescribable roar, ragged and impossibly loud, the way a wounded pit bull might howl if it outweighed a bison, and happened to be wired into the speaker system at a Led Zeppelin concert.

Below us he paused to hammer a staccato thunderclap against the ground with open palms, before circling up through the undergrowth for another pass. We held our position and braced ourselves as he came into view above us. Glowering with dark, inscrutable eyes, he screamed again and lunged forward, but didn't charge.

The staredown continued for several tense minutes, and I regained enough composure to raise my binoculars for a closer look. Though the glasses I could make out every wrinkle in his leathery face, the shadow of his brow, the dark brown eyes narrowed in anger. He had a distinctive notch running sideways from his left nostril, but I would come to know this gorilla more by his haughty, belligerent glare than any scar or nose-print.

I glanced at the trackers and they looked shaken. Suddenly I realized they had it far worse than me; I couldn't imagine how those screams must have sounded with a hangover.

The gorilla lunged again, an aggressive, stiff-legged lurch, accompanied by loud barks. We backed off, regaining our five-meter distance and coughing quietly to mimic the sound of gorillas at rest. This seemed to mollify him and he sat down, tearing at the leaves around him as if feeding.

"*Makale*," Phenny whispered, a Rukiga word meaning "Fierce One." Prunari, James and Charles all nodded their assent with wary grins, and I shakily pencilled the word into my notebook. We had named our first gorilla.

Chapter VII

America's Game

"In a very real sense, you will be a grass-roots representative of the American people while living and working in Uganda."

—*Peace Corps Uganda Handbook, 1993*

"*Agandi* John! John, how are you? HOW ARE YOU JOHN!"

"*Ndi gye*," I answered. "Fine, fine. How are youuuu?"

The kids laughed and waved back from their vantage in a field far above the roadside, continuing their chanted greetings as I walked past. In bright yellow T-shirts and folded paper hats, they stood out against the fresh-tilled soil like strange, vibrant crops, and their voices carried clear as bell tones through the morning air. People around Buhoma lived more than ninety miles from the nearest telephone, and had developed an almost uncanny ability to hold long-distance conversations, casually shouting from *shamba* to hilltop and back, as if the whole valley was wired for sound. They talked about the weather, crops, and the health of their families in a constant dialogue of hoots and calls that echoed across the hills and gave the landscape an intimate feeling, bringing the farthest fields and farms into the web of village life.

I passed a group of women walking single file down the road, arranged from largest to smallest like a line of Hungarian dolls. They balanced heavy loads on their heads: yams, passion fruit, and sweet potatoes bundled tightly in brilliant swathes of Zairian fabric.

"*Orio*, John," they greeted me.

"Yes," I smiled back and kept walking.

Most people in Buhoma still mistook me for John Dubois, my closest Peace Corps neighbor and the first *muzungu* to live for an extended time in the village. John arrived in 1991, when the Peace Corps first resumed volunteer activities in Uganda after a 19-year lapse. His original assignment involved gorillas and tourism in another national park, Mgahinga, thirty miles south in the Virunga Volcanoes. But when he arrived for his site visit, the civil war in neighboring Rwanda had spilled over the border, and shells were falling on the forested slopes where he was to live and work.

The Peace Corps quickly transferred John to Bwindi, where he pioneered my job, helping initiate the gorilla habituation process, and beginning a program to train park guides. After several months, however, he found himself more interested in community development, and slowly shifted his focus away from the forest. He began working closely with nearby schools, scout troops and a local women's group, organizing conservation education programs and fundraising activities. His enthusiasm motivated community leaders to start a new project: the construction of a campground and several local *banda*-style huts to accommodate tourists visiting the park. John secured financial assistance from the Peace Corps Small Project Fund, and extended his contract for a third year to see the enterprise through.

I first met John during training, when he visited Kajansi to tell us about his experiences working with Uganda's National Parks. He had just returned from home leave, and at first I mistook his trance-like stare at a concrete wall for the effects of jet-lag. Only later did I recognize this daze for what it was: a fine-tuned survival skill, honed to perfection through years of day-long village meetings, where people argued for hours over school fees and goats in a language he couldn't understand. His patience was famous throughout the county, and villagers loved him for it, as much as they appreciated his humor and generosity. John spoke in a loud, northeastern accent and had curly black hair with a matching mustache. I wasn't sure if people actually got us confused, or whether they simply assumed that all *muzungus* were called John. Either way, it didn't bother me. Liz had been coming to Buhoma for nearly a year, and everyone still called her John too.

I left the road and descended through a sloping, grassy compound to the park office, a tin-roofed building that housed our visitor check-in desk, two rusty filing cabinets, a shelf full of old National Geographics, and a solar-powered VHF radio, our one connection to the outside world. On sunny days, with the battery fully charged, we could talk to headquarters in Kampala, and hear other parks across the country calling out to each other in bursts of static and garbled voices. But when the clouds set in, the airwaves grew quiet, and we had a hard time even raising Ruhija, a ranger station twenty miles away. There are a lot of cloudy days in a rainforest, and at times we felt cut off, like some forgotten frontier town far beyond the range of railroads and telegraph. It came as no real surprise when a television show called "The Ends of The Earth" inquired about filming an episode in Bwindi.

This morning, however, the radio had a good charge, squawking away unnoticed in a corner as the guides and trackers gathered for another day in the forest. "*Oriaregye*," they called, and "*Agandi, Tour!*"–using the distinctive Rukiga pronunciation of my name. But at least they knew who I was, and I paused to greet everyone before circling around the back of the building to find John.

He lived in a single room, separated from the office by a thin wall that stopped four feet shy of the ceiling. When other volunteers reminisced about the scary depths of Peace Corps housing, they talked about John's place, a shadowy realm of overheard office chatter and radio static, cluttered with laundry, moldy food, half-painted campground signs and, oddly, baseball equipment.

In 1992, the Little League of America began a campaign to spread the gospel of baseball around the world. They contacted hundreds of Peace Corps volunteers, and began shipping equipment to remote villages and communities.

"I thought they might send me some balls and a couple of bats," John told me. But several months after he signed on as Southwest Uganda's Little League Chairman, the crates started arriving. And they kept coming. Box after box of top-quality gloves, balls, and bats, with plates, catcher's masks, helmets, run-counters, and batting tees—enough gear for an entire league of well-supplied teams. It took up half the room and left John living in a cramped corner of open space near the door.

"It must be fifteen thousand dollars worth of stuff," he said, laughing at the heaps of mouldering boxes. There was more equipment waiting in Kampala too, stockpiled for the day that baseball supplanted soccer as Uganda's national sport. If John's experience was any indication, however, that day was a long way off.

"I have to make everyone wear batting helmets all the time," he explained. "It's just too dangerous otherwise." For most soccer-bred Ugandans, the concept of throwing and catching a ball with your hands was completely foreign. I watched John trying to teach the basics at a Boy Scout meeting: thirty barefoot kids in ill-fitting helmets with baseballs flying all over the yard, whapping off people's heads and bouncing out of sight into the shrubs. "This may be hopeless," John confessed.

Luckily, coaching baseball played a relatively insignificant role in John's overall agenda. His work took him to villages and community groups throughout the area, and today I was joining him for a trip to Butagota, the largest town in the area, and the site of a weekly market. John needed to visit several primary schools and deliver newsletters to their fledgling Wildlife Clubs. I had a more fundamental goal: food.

My first weeks in Buhoma had flown by in a blur of apes and rain, and long days in the forest. The bread, rice, tomatoes, and other exotics I'd brought from Kajansi had been quickly devoured, or gone moldy in the damp forest air. It was time to stock up on local staples: beans, bananas and the small, yellow-skinned potatoes known as "Irish." My Norwegian relatives were willing to overlook this geographical slight, and took great comfort in the fact that I'd been stationed in Uganda's main spud-growing region.

"Do they have potatoes?" my aging great-uncle had asked, his one concern on the day of my departure. "Yes? Then you'll be just fine."

John and I split a pineapple for breakfast and set off down the road to Butagota, following a labyrinth of footpath shortcuts through the village* hills. The three hour hike stretched into four as we stopped to chat with people along the trail. They asked about the weather in Buhoma, the state of the park, and what news we heard from America. Our stops grew more frequent as we neared the town, where pedestrian traffic swelled to a steady stream.

Market day in rural Uganda highlights an otherwise unchanging week. In Butagota, merchants gathered from around the region, and people flooded the town to socialize, drink, or trade the extra produce from their *shambas*. We arrived with a lively throng of people from Kanyashande, everyone but us bringing something to sell: peanuts and papaya, chickens, clay pots, and a herd of young goats, braying and pulling at their bark-rope tethers.

Lying less than a mile from the Zairian border, Butagota's prosperity depended largely on illegal trade, coffee smuggling, and gold. There was a tea factory and shops lining both sides of the road, but the wealthiest merchants all had ties to the border trade. John and I stopped for tea with one of the town's leading citizens, a heavy, confident-looking man named John Nkunda.

---

*   Most Ugandan villages have only a small trading center, or no discernable locus at all. The word 'village' usually denotes a small geographical area, and all the farms and families therein.

"He's involved in just about everything around here," John whispered as we walked through a courtyard piled high with sacks of dried coffee.

Nkunda owned the only lorry in the area, and charged a steep fee to transport goods or building materials. I waited in the sitting room, chatting about coffee with one of his sons, while he and John negotiated over a load of cement for the Buhoma Community Campground.

John was still shaking his head over the deal on our way to the market, "I can't believe he's charging us so much. The transportation costs more than the cement."

Near the center of town, construction was well underway on the brick and mortar building that would house Butagota's new tourist lodge. It was the town's first effort to capitalize on Bwindi's growing tourism market, and could provide stiff competition with the Buhoma campground. When we passed, a man was balanced high up on the rickety scaffolding, painting something over the entryway in large red letters: "Nkunda & Sons." Ah. John didn't say another word about the price of cement.

We followed the crowds down a slippery side trail to the market itself, and quickly became separated in the shuffle. I found myself suddenly alone with my shopping list in a sloping, muddy field cluttered with lean-to stalls, where bustling mobs of people argued and haggled over a chaotic array of baskets and fruit, green tomatoes, used clothes, buckets, spoons, goats, tin plates, bicycles, hardware, fabric, cabbages, kerosene, sugarcane, chickens, cassava, *matoke* bananas and black rubber sandals. I entered the melee and immediately drew a crowd that followed me for over an hour, hooting with laughter and crying "*Muzungu, arumanya!*" (He knows Rukiga!) as I struggled to remember my market survival phrases: *ebi-himba*—'beans;' *akavera*—'plastic bag;' *Nooseera*—'you're ripping me off.'

I found John rooting through the piles of used clothes.

"You can get some great shirts here," he said, holding up a zippered jersey in burgundy velour. Fashion gems from every decade since the 1950s are still available at rural Ugandan markets. When charities in the States and Europe send loads of cast-off clothing to Africa, they fuel a burgeoning and lucrative second-hand industry. I once saw a shipment of shoes arrive at Kampala's Owino market, a major distribution center. Wholesalers lunged forward in a rush, shouting and grabbing pumps and loafers like desperate traders in a stock market crash.

Durable, brightly-colored clothes were always in high demand, regardless of their strange foreign logos. This system led to a number of eye-turning combinations, like the old woman in Mbarara wearing a tie-dyed shirt emblazoned with marijuana leaves, or the burly Kampala street thug in a pink "Sexy Grandma" hat. Butagota's selection offered everything from Levis and Reebok sweat pants to polyester suits, a Denver Broncos parka, and two 'Dukes of Hazard' T-shirts.

The sky was dark with coming rain when John and I finished our shopping and headed for the first school. Two hundred shouting children raced across the yard to meet us, and we joined in a mass soccer game, running over the lumpy field and kicking a makeshift ball of tightly-bound banana fibers. The teachers had all gone home for lunch, and John sent someone to find them as the first heavy raindrops pelted down. We crowded into a tiny, brick-walled school room and the sky opened up, thundering a staccato deluge onto the thin metal roof.

The school had a wooden blackboard, but no chalk. There were no books in the room, no desks and no pencils, and most of the kids had nothing to write on. Although the government listed education as a top priority, few small towns were better off than Butagota, and village schools were worse. Parents struggled to pay tuition and fund improvements, but rural children, particularly girls, seldom made it beyond the primary level.

The kids danced and sang local songs to pass the time, clapping and chanting in high, piping voices over the storm's deafening rhythm. Gathered close to us in the shadowy room, their faces mingled into a shifting tumult of wide brown eyes, bright with smiles and music. The rain continued and the teachers never came back, but there was something potent, even hopeful, in that simple chorus, and it was impossible not to sing along.

John led a round of "Row, Row Your Boat," and then we tried to teach them "Doo Waa Diddy," without much success. When the storm passed, we moved outside and the kids followed us back across the playfield, shouting and waving as we made our way down the road. They needed far more than songs and soccer, but the best community development is subtle work. John and I accomplished two thirds of the Peace Corps mission by simply being there, sharing something of our own culture and learning something from theirs. To do more is often to presume too much, I reminded myself, like catcher's mitts and batting helmets for a town that has no textbooks.

Chapter VIII

They're Everywhere

---▼---

"The pismire, known to the people as the 'chungufundo,' is a horse-ant, about an inch in length, whose bull-dog-like head and powerful mandibles enable it to destroy rats and mice, lizards and snakes. It loves damp places...It knows neither fear nor sense of fatigue; it rushes to annihilation without hesitating, and it cannot be expelled from a hut except by fire or boiling water. Its bite...burns like the pinch of a red hot needle... and it may be pulled in two without relaxing its hold."
—Sir Richard Francis Burton meets safari ants, *1857*

The bride wept, rocking back and forth in her chair and moaning rhythmically in abject misery. Occasionally, an aunt or one of her sisters leaned over to console her with soothing whispers, or a pat on the arm. But everyone else at the party seemed to ignore her sorrow, particularly the groom, who stood outside with his brothers and friends, joking, and drinking liberally from a calabash of sour banana beer. It was a perfect wedding.

Dominico rushed about the compound, greeting guests and helping the women serve up steaming platters of boiled goat, cabbage, *matoke* and

'Irish.' He swept past me, stopping to refill my mug from a brimming vat of *tonto*.

"*Webale kwija, Ssebo,*" he crowed happily, thanking me again for coming. When he first mentioned his son's wedding, I envisioned stopping by for an hour or so after work, but the celebration had been raging all afternoon, and now, after several cups of *tonto*, the slippery trail downhill to my house was looking less and less appealing, perhaps impossible. I took another sip of the smoky, bitter brew, and couldn't help thinking of Tom, a true aficionado of local hooch.

The Ntales had thrown a huge going away party on my final night in Kajansi, complete with speeches, feasting, and endless gallons of pineapple wine. Tom invited every Peace Corps Volunteer he'd met, and we danced for hours on the lawn, swaying to disco, Elvis, and Dolly Parton. All of Annette's regulars were there, and I said my goodbyes to most of the neighborhood on the dance floor, a fitting close to the social frenzy of my months with Tom. The next morning he packed a small jerrycan of left-over *munanasi* into my luggage as a parting gift.

"You will need it there," he told me, adding in near-disbelief, "Those Bakiga don't make any wine. All those pineapples they grow—for nothing!"

Tribes in Uganda relish teasing each other over food preferences, and a proud Baganda will scornfully dismiss the entire western half of the country as "millet eaters." But in the case of *munanasi* vs. *tonto*, Tom definitely had a point. I learned how to make *munanasi* at Annette's place, with a big pineapple grater and a series of shiny aluminum pots. After three days of fermentation, the final product resembled a thick, tart cider. Brewing *tonto*, on the other hand, involves a bunch of barefoot men stomping around in a hollow log full of bananas. In the end, it's tough to say whether it tastes more like the bananas or the feet.

The evening stretched towards darkness and thunder rumbled overhead, mingling in subtle counterpoint with the frantic cadence of Dominico's two musicians. They stood together in the downpour, hammering with mallets on a large wooden drum stretched tight with cowhide. Water sprayed up from the wet leather with every beat, surrounding them in their own uprisings of tiny rain, like a pulsating, syncopated fountain.

An informal chorus of women sang along, but dancing had been postponed until after the storm. We all crowded together under a makeshift shelter of reed poles and banana leaves, talking, eating, or just sitting back to watch the celebration unfold. Rain leaked steadily through the scant roofing, and soon the compound was a morass of mud, trampled into pools by hundreds of sandaled and barefoot revelers.

During lulls in the conversation, I could hear the neglected bride chanting something through her sobs. The words were indistinct, but it must have been her variation of the traditional nuptial lament:

"I go to distant countries,
to other people's homes where people never visit
I will see you no longer: goodbye, goodnight;
bid farewell to my mother."

Bakiga custom requires every newly-wed woman to mourn publicly over leaving her family. The tears and misery should last throughout the ceremony, and sometimes for days afterward. Even Western-style, city marriages follow this ritual; to appear excited about the marriage, or even marginally cheerful, would be a great affront to one's parents.

Of course, in many cases the melancholy need hardly be acted. Families often arrange marriages without the daughter's consent, and young girls may find themselves paired with elderly men or complete strangers. Ugandan matchmakers place far more importance on haggling over the bride price, than in creating wedded bliss. The cost of a wife depends largely on the wealth (or perceived wealth) of the groom's family, but the fees are always exorbitant. In Buhoma, young men complained constantly about their 'dowries,' and how to pay them off. The sum usually included between one and two hundred dollars in cash, supplemented with at least twenty goats, ten cows and quantities of *tonto*, millet and other commodities—far more than anyone could afford to pay at one time. Remittance often stretched over years, or even decades, forming a complex network of debts between families and clans throughout the area.

The bride price system plays an important cultural role in binding Bakiga communities together, but at significant cost to the social status of women. The hardship of long-term payment makes men more likely to treat their wives as property, particularly second or third spouses. As Agaba Philman, a porter who worked for John, once told me: "For twenty goats, she will wash my feet!"

Women carry out the brunt of agricultural and household labor for every Ugandan family, forming the backbone of village economies throughout the country. Local and international development groups recognize this fact, and women's rights are slowly improving, but the pace of change slackens as you range further from the influence of urban centers. Still, Dominico's young daughter-in-law would have had far more to cry about in previous times, when meeting the male members of her new family involved sitting on a wooden stool puddled with their combined urine. After this symbolic ritual, any of them were free to demand sex.

"If a man came home to find his brother's spear beside his door, it meant he was with the wife," Enos Komunda once told me. "He could either wait, or go directly to plant his spear at the brother's house."

A former park guide with a strident voice and natural flair for story-telling, Enos was in his first year at Kabale's Protestant seminary. We missed him at the park, and John and I both had qualms about his new career choice. Corruption in the Church of Uganda allowed high-level clergy to live in luxury from the offerings of poor rural parishioners, and a suspicious number of their 'local development' projects never made it beyond the fundraising stage. Still, we had encouraged Enos to go, and even agreed to help pay the school fees. Village life offered few opportunities for advancement, and we couldn't begrudge him his chance for an education and a good job in the ministry.

For his independent research, Enos chose to study local religious and cultural traditions. He spent his school breaks interviewing elders around Buhoma, and we often talked at length about various Bakiga customs.

"For a woman, it was taboo to eat chicken," he told me seriously one day, as we bounced along towards Kabale in the back of Liz's car.

"Sure?" This sounded interesting, and I prodded him for more. "Why? Was there a belief it could make them sick?"

"No, no," Enos barked a laugh and shrugged. "It is because we men wanted to eat all the chickens!"

I laughed with him, and he continued.

"The same was true for goats. Girls were allowed, but when they reached a certain age, it became taboo. You know these intestines?" He made a braiding motion with his hands. "The sweet ones?"

I nodded. Throughout the region, tripe is considered a delicacy, and enterprising cooks braid the entrails into long ropy strands for a special stew.

"When a girl became too old, they filled the intestines with *empazi*. She bites it, the ants come out, and she never eats goat again!"

Thinking of it now, I shuddered and discreetly poked through my bowl of stew, relieved that Dominico hadn't served me anything braided. Also known as 'safari' or 'fire' ants, *empazi* travel in unstoppable, mile-long columns that have earned them a more sweeping title in the folklore of tropical Africa: army ants. One misstep in the forest and they swarm up your legs in an instant, biting simultaneously from a hundred different points, like tiny stabs from a rain of hot nails. More than once I'd seen guides, trackers, and even tourists frantically strip off their clothes to shake out the writhing hordes. Cultural symbolism aside, chomping down on a mouthful of *empazi* would be enough to spoil anyone's appetite for goat. Or chicken. Or chocolate, French fries, lasagne, doughnuts—anything! In the States, army ants could make someone's 'Info-mercial' fortune as a dieting aid.

Dominico invited me into his house for dinner, to a private room with a kerosene lamp, a table, and a setting for one. As a *muzungu*, people in the village often treated me with a kind of habitual deference left over from colonial times. My position of authority with the park added to their preconceptions, and many locals were shocked to see me engage in physical labor, or eat from a common bowl. At social gatherings like Dominico's, the host usually singled me out for special treatment, an honor that was complicated and awkward to refuse. Instead, I simply extended the invitation to other people at the party: two porters from the trail crew and Caleb Tusiime, a park guide who was training to become a nurse.

We ate quickly, talking and laughing by the lantern's ruddy orange light. Someone called for another jug of *tonto*, and our conversation soon turned (as it would with any group of young bachelors in Buhoma) to the topic of dowries. I teased them, naming women from the village, and asking "*Mbuzi zingahi?*"—'how many goats?' When I mentioned that we paid no bride price in the States, they shook their heads in envious disbelief. "And", I added, "the woman's family even pays for the wedding party."

"Is it?" Caleb's laugh was high and drawn out, like the shout of a loon. "In America, I think I would be married many times!"

Later that night the clouds finally parted, ghosting away from a crescent moon like giant, flat-bottomed dhows. Walking home proved easy in spite of the *tonto*, with my path lit clear as midday by the glow of moon and stars. I paused outside my house, and listened to the quilted echo of Dominico's drums spill out across the valley. The rhythm was almost visible, a half-imagined mist, murmuring with cricket rasp as it drifted over the farms and treetops. I heard a wood owl call, and for an instant the forest seemed to resonate with silver-blue sound, wet leaves throwing back moonlight like the pale shine of a thousand-thousand dimes.

((((

Several days later, the trackers and I passed Dominico's *shamba* on our way to the forest. He waved to us wearily, squinting through the smoke of his tiny wooden pipe, as if still hungover from the party. Next-door, we found his neighbor, Kazungu, preparing beer for another village celebration.

"*Agandi, baa Ssebo*," he greeted us, offering up a sample of his labors. We paused while he scooped a cup of brew straight from the hollow log where his youngest son was trampling ripe bananas. Draining off the frothy pulp, everyone tasted from the cup with thanks and nods of approval. Unfermented *tonto* is a sweet and surprisingly palatable juice called *omubisi*, favored by children and non-drinkers. We filled a plastic jug to take with our lunch and set off again, climbing rapidly towards the forest edge.

Katendegyere group's trail led us along the steep, southeastern flanks of Rushura, ascending gradually through a dim realm of canopied shade and pale branches, hanging down with vines. We found fresh sign and night nests near the Zairian border, and Charles bent down to examine the dung.

He flattened two piles with the toe of his boot, filling the air with a pungent, barnyard reek. Pointing out certain patterns of thin fiber in the bolus, he turned to me and shook his head with a rueful smile. "Again they are chewing bananas."

We followed the gorillas' trail into a steep-sided valley recently cleared for cultivation by Zairian farmers. Much of Katendegyere group's former range had eroded over the past decade, as forested areas outside the park fell prey to the axe, and the increasing demand for agricultural land. Nearly everything on the Zaire side had been cut, and when we found the gorillas here, they were usually raiding banana *shambas*, making best use of the strange new trees that had suddenly sprung up in their back yard.

Today was no different, and from our vantage on the hillside we could see the gorillas clearly. Their dark shapes looked bulky and out of place in the fields, moving slowly between broad-leaved banana plants.

As we made our way down the slope, a farmer called out from the smoky doorway of his hut. We stopped to talk and he came across the compound to meet us, reed-thin and barefoot, with proud, aristocratic eyes that belied his tattered clothing and ramshackle home. One of our favorite short-cuts through the forest crossed this *shamba*, and on other days he had greeted us with gifts of passion fruit or *omubisi*. But now his eyes were cold, and he raised his voice in an angry shout: "Your animals, they are bleeding us! What will we eat now?"

I tried to reassure the man, but knew that compensation for his loss would be a long time in coming. The park and IGCP offered a small 'token of appreciation' to farmers with gorilla-damaged crops, but the system was new, and we didn't have permission from Zairian authorities to compensate people on their side of the border. Mishana James stayed behind to explain things, while Charles, Prunari and I descended the hill and positioned ourselves near the feeding apes.

We watched the group's second silverback reach up with one huge arm and pull down a 15-foot tree. I cautiously snapped pictures, but he seemed indifferent to the camera, glancing our way with placid eyes, and scratching absently at the long black fur on his shoulder. Then, with a casual flex and a sideways tear of his teeth, he separated the trunk into ropy strands and began chewing noisily.

Bananas are the only local crop favored by gorillas, but contrary to popular belief, they rarely eat the fruit itself. The cartoon image of an ape peeling fistfuls of yellow fruit is the product of zoos and circuses, where caged animals developed a taste for whatever their keepers fed them. In the wild, it's actually the watery core of the plant stem that draws them from their forest home. Stands of banana trees like this one, planted near the forest edge, may be destroyed long before the farmer ever has a chance to harvest.

We sat quietly while the silverback wrenched down another tree. He positioned himself between the two fallen stems like a choosy diner in a buffet line, alternately yanking mouthfuls of wet fiber from one and the other. Prunari leaned forward and whispered a name in my ear, "Mutesi," —the lazy, spoiled child. I hastily sketched his noseprint while occasional belches and rustling leaves revealed the presence of other apes resting and feeding around us. We couldn't see them, but we knew who was there: Mugurusi, "Old Man," the shy, lead silverback; Nyabutono, "Little Lady," his constant companion; or Karema, "Cripple," the calm young female with maimed fingers. Our sightings of the group had improved dramatically in recent weeks, and I felt we were coming to know this family of apes by their personalities as well as their physical traits.

So when we heard a series of sharp pig-grunts in the tall shrubs beside us, we looked at each other with immediate smiles, "*Makale.*"

Pugnacious, and slightly too young to challenge the group's three adult silverbacks, Makale was a growing, black-backed male, and the group's self-appointed watchdog. He served as our barometer for judging the group's mood, a surly pressure vent for any kind of tension or collective nerves. But while Makale's grouchy personality had become almost endearing, his habit of screaming and charging to within inches of us exposed him—and the rest of the group—to a variety of dangerous human diseases. Gorillas share ninety-eight percent of the same genes as *Homo sapiens*, and can succumb to the same viruses and bacteria that make us ill, as well as many more that we carry unknowingly. We followed the park's health and minimum-distance regulations with vigilance, but convincing a charging gorilla to observe the same rules is another matter altogether. Today, however, feeding on a large banana stem demanded all of Makale's attention, and we didn't see more than the dark shadow of his glare, deep within the bracken.

The farmer above us had returned to work, swinging his hoe at the earth and turning back the soil for a row of young banana shoots, as if in a conscious race to offset every tree the gorillas had devoured. This convergence of forest, wildlife, and agriculture represented a true microcosm of conservation problems in the region, where a rapidly growing human population is pressed right up to the edge of protected areas. Although fields like this one had been vital gorilla habitat five years ago, it's difficult to blame the loss on any individual. People living in such a remote valley were among the poorest in Central Africa, struggling at the lowest levels of subsistence. Of course the farmer was angry; any crop damage on this steep, marginal land meant an immediate loss of food for his family.

Gaining the support of local people presents one of the fundamental long-term challenges for Bwindi Impenetrable Forest. When the park was created in 1991, thousands of villagers found themselves suddenly cut off from a traditional source of bush meat, firewood, timber, and medicinal herbs—their customary habits stopped for the sake of international tourists and an animal who raided their crops. Starting from this setback, park managers attacked the public relations problem with a combination of experimental policies.

Communities surrounding the forest were invited to elect representatives to a park advisory committee, ensuring local input in major management decisions. Their suggestions helped the park establish 'multiple use zones' along the forest periphery, and begin licensing designated herbalists from each village to organize a controlled, sustainable harvest. Beekeepers were also given limited access, and certain communities resumed the collection of rafia palms for traditional weaving and basketry. Many of these programs were administered with the help of Development Through Conservation (DTC), a CARE International program working on conservation education, family planning, and agricultural development in communities bordering Bwindi and Mgahinga.

The multiple use program went ahead with relative ease, but controversy surrounded the centerpiece of Bwindi's community-development efforts, a plan to distribute portions of tourism-generated revenue directly to local villages. Communities would submit funding proposals for their own projects, avoiding the typical 'top-down' approach of government-sponsored development. Early ideas in Buhoma ranged from a new roof on the primary school, to road improvements, or equipment for the medical clinic.

Most people supported the revenue sharing concept: helping local people associate gorillas and forest conservation with a tangible increase in their standard of living. Contention arose over the two major points of implementation—how much to share, and how to share it. Could a park system that already relied on foreign aid to balance its books really afford to give anything to villagers? Conservationists had watched similar programs struggle in Kenya, where project after project failed due to the logistics and politics of distributing money at a local level. Bwindi's program hoped to avoid these pitfalls, ensuring accountability by allowing community representatives to choose which of their own projects to fund. Additionally, the World Bank would share administrative costs and guarantee a steady income through its new Bwindi Trust, a special account to support park management, research, and community development in perpetuity. In any form, revenue sharing represented a new concept for Uganda's national parks. Bwindi's attempt was the pilot study to determine whether headquarters would implement a country-wide program.

Before heading home, we advanced slowly towards the gorillas, trying to coax them back into the forest. After such a calm day, I hated to risk disturbing the group, but leaving them alone in the fields only invited worse treatment. When we were gone, the farmers would begin to shout, beat on pans and throw rocks—a good technique for driving gorillas from your *shamba*, but a serious setback to our habituation efforts.

Veils of rain drifted over us, and most of the gorillas had already retreated into a bower of heavy shrubs, but Mutesi was still feeding as we approached. He stared at us warily and began chewing faster, the pithy juice running sloppily down his chin. He was obviously nervous, but still appeared to relish every bite. For Katendegyere group, the sheer epicurean pleasure of a fresh banana stem seemed to outweigh any risks involved in raiding the *shamba*. I doubt that even the old Bakiga 'stuff the food with *empazi*' trick would have slowed them down. Mutesi certainly looked like it would take more than a mouthful of army ants to move him from his feast. But we kept edging forward and finally he gave way, disappearing into the forest with a tattered banana tree dragging from his fist, like the last minute spoils of a Viking raid.

That evening, a late cloudburst hammered the eaves of my house, drumming against the banana-thatch like endless padded applause. I went to bed early, listening to raindrops leak through the roof and rattle into my rag-tag collection of pots and basins in a cheap, tin echo of the fading storm. If the rain continued, I knew the park staff would be short-tempered in the morning. No one in Buhoma slept well when they spent half the night shifting their bed around the house, trying to avoid the constant indoor drizzle of a sodden thatched roof. Ephraim and James Kawermerwa had recently patched my bedroom ceiling, and I felt a little smug as I drifted off in leak-free comfort.

Sometime later, a new sound in the house brought me slowly back to consciousness. The rain had stopped, but there was something else, a subtle noise pervading the blackness around me. I listened carefully as I came awake, but couldn't pinpoint the source. It seemed to come from all sides, and sounded like...seething. *It couldn't be ants*, I told myself with a nervous mental chuckle. *You can't hear ants. Go back to sleep.* But the skittering racket continued to grow, and rest was impossible until I settled the mystery.

Lighting a match in the rainforest wasn't always a simple task, but it had never seemed to take longer as I scratched through half a box of damp sticks. Finally, a light flared up, and I glimpsed something that made me wish I'd never opened my eyes: *empazi,* millions of them, blanketing the walls, covering the floors, dropping from the ceiling, and swarming up the bedspread towards me in a teeming mass. Then the match went out.

It was a classic horror show image in real life, and I reacted with the same 'let's-leave-the-weapons-here-and-split-up' logic that dooms so many movie characters. Without even pausing to think, I let out a yell, grabbed the nearest lantern, and leapt out the window in my underwear.

Dashing across the wet lawn, I frantically pulled ants from my hair and swept handfuls of them from my skin, making a desperate attempt to stay calm. I thought of the gorillas, and how they would simply move their nests if a trail of ants invaded during the night. No problem. With Liz in Kampala, I had an empty, fully-furnished house right next door. The *empazi* blitzkrieg in my bedroom would probably look a whole lot better after a good night's rest. If I was lucky, the army would have moved on by morning.

I walked across the yard, shivering in the misty cool night, and shaking my head. People in the village already thought I was slightly crazy, living alone so close to the forest. *They should see me now*, I laughed to myself as I leaned forward to unlock the door. But all humor at my predicament disappeared in an instant, when the dark, shadowy wall of Liz's house began to writhe, and a new wave of ants rushed up my bare arm. This is when panic truly set in. Dropping the lantern, I pawed spastically at my arms and face, and found myself suddenly screaming at the top of my lungs, completely out of control: "THEY'RE EVERYWHERE!!!"

Ten minutes later I was huddled in the pit latrine, the only ant-free structure on our whole hilltop compound. I might have spent the night there if I'd thought to put on some clothes, but even Africa gets chilly in a pair of boxer shorts. When I regained some sense of composure, I returned to survey the situation. My house was a complete loss. The invaders commanded floor, wall, and ceiling space in every room, centering their attack on the kitchen, where I glimpsed ants pouring out of the cupboards and swarming over a bin of dirty dishes in a dense, almost liquid mass. Occasionally, a cricket, spider, or small lizard would fling itself down from the rafters, twitching spasmodically under a living blanket of tiny, voracious attackers.

I grabbed a sweater and raced back to Liz's house, where one chair and a coffee table seemed to have escaped the *empazi's* notice. More importantly, I knew Liz's cupboard contained two items crucial to my newly-formed defense plans: a can of Doom bug spray and an unopened bottle of cognac. I used liberal amounts of both as the night wore on, and by morning, my mood placed me far above the leaky-roof crowd, mournful brides, and even Makale, as the most ill-tempered resident of Bwindi Impenetrable Forest.

Chapter IX

Arming Makale

▼

"He was evidently one of the largest sized gorillas. In the gloom of the forest he appeared to us to be above six feet high. His jet black visage was working with an expression of rage that was fearfully satanic. His eyes glared horribly...As it took the next step and appeared about to spring, Jack pulled the trigger. The cap alone exploded!...Almost as quick as thought Jack hurled his piece at the brute with a force which seemed to me irresistible. The butt struck it full in the chest, but the rifle was instantly caught in its iron gripe. At that moment Peterkin fired, and the gorilla dropped like a stone...not, however, before it had broken Jack's rifle across, and twisted the barrel as if it had been merely a pipe cleaner!"

—R. M. Ballantyne
The Gorilla Hunters, 1890

A change of seasons in Bwindi brought subtle, but welcome variation to our daily allotments of heat and rain. January marked the beginning of the dry months, and the frequency of showers slowly tapered off until whole days might pass without a single shower. The heavy tropical sun hammered red dust from the hillsides, and smoke from wildfires hung high in the air like a pall, muting the landscape to an understatement of tree shapes, valleys, and hazy green hills. Hiking became an exercise in sweat, with the swelter of midday weighing at my shoulders like a leaden cloak, and after a few weeks I found myself longing for the rains again. I asked Ephraim if the dry season was always so hot.

"Sure. It's too hot," he affirmed solemnly with a shake of his head. We were sitting outside together, huddled in the tiny shade beside my house.

"But then the wet season comes: too much rain," added Betunga William with a laugh. He was lying in the grass with a shirt over his head. "You can see we are suffering."

Two hornbills rose in a brief lazy glide over the forest, as if testing the air. Their huge pied bills looked heavy as iron, and their drunken caws sounded thin and distant, muffled by sunlight.

"Yes," I agreed, smiling. "We are suffering here together."

Our suffering aside, the days of sunshine brought with them a new sensation to the forest: *dry*. Long-sodden laundry finally lost its musty smell, and the patches of mildew under my bed began to recede. The ceiling ceased its constant indoor drizzle, and I regained the use of pots and basins lost for months to permanent leak collection. People everywhere took the opportunity to patch their roofs, and the price of good banana fibers jumped 300 shillings in the village.

In the afternoon, clouds still massed and threatened, but any rain was usually brief and quickly swallowed up by the dry soil. Mornings dawned endlessly clear, with a chorus of forest birds singing upwards to a dull blue sky slung low between hilltop horizons of green.

Mountain gorillas apparently use the dry season to train for marathons and long-distance orienteering courses. Katendegyere group normally moved less than a mile a day, but in the dry months they could travel twice as far, climbing and descending a circuitous path through the steep-sided valleys. Their wanderlust didn't stem from a sudden desire to explore the forest, and they weren't necessarily trying to challenge their human pursuers.

The motivation for their journeys was probably rooted in a far more simple impulse, thirst.

Gorillas rarely drink free-standing water, relying instead upon the moisture in their diet of leaves, bark and fruits, and the raindrops they lick from their fur during heavy showers. In the dry season, however, fewer plants produce fresh greenery, and the apes must range further to ensure a steady supply of moist vegetation. They sought out the new growth in clearings and in the dense thickets along the forest edge. While they might linger in large clearings like the burn on Rukubira hill, they often crossed long stretches of closed forest to visit several feeding sites in a single day.

In the thick tangle of forest glades, gorillas trample vegetation with the subtlety of small steamrollers, and their paths are easy to follow, but when a group travels quickly through closed-canopy forest where the undergrowth is sparse, their passing may displace only a few scattered leaves and twigs. It's worse in the dry months, when their feet leave no prints and the vegetation springs back into place unaltered. At such times the trackers would usually split up, circling until one found a telltale sign: fresh dung, misplaced leaves, or a sapling stripped of its sweet green bark.

Prunari and Charles divided to widen their search on one such occasion, while I waited where we had lost the trail. In a dry forest I couldn't track my way out of a paper sack, and knew when to let the professionals work undisturbed. I watched them moving further into the shadows and began rooting for Charles, who had taken the easy, downhill path. But I had little hope; when mountain gorillas have a choice, they always go up.

It was midafternoon and the forest lay dim under the shade of thunderheads. The gorillas had circumnavigated the entire Muzabajiro river valley, and led us to this point near the top of Rushura hill. Their general direction pointed to the forest edge and we expected to find them raiding banana plantations, or feeding in the tangled stands of bracken just outside the park.

I heard Charles give a long, questioning whistle from below, answered by a quick affirmative from up the slope. Sure enough, Prunari had found the trail again and I heard the metallic ring of his *panga* as he began hacking through thick undergrowth near the park boundary. Charles and I followed up the steep hill and rejoined him as we emerged, squinting, into the full light of day.

The forest edge stretched away to the south in a near-straight line separating jungle from small farms along the Zairian border. Below us, cultivated fields descended in serpentine undulations to the broad flat plain of the Great Rift Valley. The immensity of the view was framed in the distance by a blurred shadow of the western rift escarpment rising up in Zaire. On clear days the Virunga Volcanos would loom like distant pyramids, but now they were lost in clouds and dry season murk. The hillsides were blackened in patches from brushfires, and plumes of smoke rose slowly from the valley floor like hundreds of tiny reverse rainstorms, spilling upwards to a sea of bluish haze.

The dry season in Uganda is a time for burning. Throughout the region, farmers set their fields ablaze, spurring a growth of new grass for their livestock, and clearing the ground for planting. In savanna landscapes, fire has played an important historical role in maintaining a balance between open plains and woodlands over time. Agricultural areas are more sensitive. Uncontrolled fires can actually damage the topsoil, and the loss of ground cover leads to increased erosion, but burning is still a popular farming tool. It provides a quick boost of fertilizing ash for young shoots, and is by far the easiest way to clear new or overgrown croplands.

Around Bwindi, selected fields like those below us were burned to prepare the ground for millet, a staple food and culturally important crop for the Bakiga people. Widespread African staples like bananas, potatoes, rice and maize are all relative newcomers to the local diet, but millet has grown here for thousands of years, a hearty grain that thrives in the rugged hill country of southwest Uganda. Within two months the charred ground would be lush with knee-high stems of a green so pure it seemed to glow from within. Fields of young millet stand out against the cultivated hillsides like vibrant windows into an underworld of primal verdure.

At harvest time women gathered in their family plots to pull up dry stalks, beating the red grain free with long-handled brooms and branches. Crushed and mingled with cassava flour, millet is served as a pasty starch called *oburo*, a vital dish at any wedding, or holiday feast. Families store the dried grain in pots or large baskets to ensure a ready supply for any significant gathering, and young brides must make a gift of millet to their in-laws before they can be fully accepted by the family. Any surplus may be cooked into a watery sweet porridge, or mixed with sorghum and fermented into a popular variation of local beer.

"A man cannot dig millet," Enos Komunda once told me. "Bananas, yes. Potatoes, yes. Even maize. But if a man touches millet, it will be sour," he explained, spitting into the earth for emphasis.

Watching the women toil and sweat in the steep hillside fields, I had a fairly good idea how that particular tradition was born. Bakiga men have a remarkable flair for keeping themselves on the feasting and beer side of agriculture.

Prunari moved ahead again, tracking the gorillas' path into a tangled thicket. We followed carefully. In thick vegetation, a sleeping gorilla is just another shadow, and more than once we had nearly stumbled into the midst of the group during their midday rest. Today, however, they were hungry after their long journey through the forest, and we soon heard them feeding nearby.

Mugurusi barked in mild alarm at our approach and we grunted a calm response, stopping about twenty feet from the shaking leaves. The group was invisible in the tall shrubs, but we could hear them belching and chewing around us, and glimpsed the occasional arm reaching up to grab at particular branches.

We settled down in the leafy shade and Prunari fell instantly asleep, while Charles pulled off his boots to adjust his tattered socks. I sat with my binoculars and notebook ready, trying to identify individuals in the group from brief flashes of black through the foliage.

Clouds of insects hummed around me in the heat, landing and taking flight from my bare arms in a constant flurry, like tiny, living dust storms. I waved them away halfheartedly. In Africa, one soon reaches an impasse with flies. To fight them is a quickly-learned lesson in futility, so you simply grow accustomed to their constant feathery touch. Watching one crawl slowly across the back of my hand I realized with a start that these weren't flies at all, but perfectly formed black bees the size of a pencil point.

Bwindi Forest harbors an immense and largely unstudied insect fauna, but I knew there were at least six different species of tiny black bees. Stingless and docile, they were much favored by the Batwa pygmies as a source of medicinal honey. Dominic, a graduate student from Makerere University, had studied the stingless bees of Bwindi for his Master's thesis. He told me of searching the forest for hives with a Batwa pygmy guide. They came into a small clearing and the guide held up his hand for silence. From a cluster of white flowers, one stingless bee flew up and disappeared into the forest.

"We wait," the man said, and Dominic sat with him near the flowering shrub. Twenty minutes later the bee returned and they watched it feed from the blossoms before it disappeared once again into the woods. Dominic sat mystified as they waited for the bee to make one more trip into the clearing. When it was gone the guide rose and they walked for half a kilometer through the trees, straight to the hive.

This story impressed me. They say that the Batwa developed their uncanny knowledge of the forest from generations of living and hunting in its shady depths. Dominic's guide recognized the species of bee and knew the type of rotten log where it was likely to nest. From its trajectory in the clearing, and the time it took to make a round trip, he estimated the location of the hive. Simple.

I watched a bee fly up from my arm. It's shiny black body vanished instantly into the leaves, a single ebony pinprick in a shifting curtain of green. My respect for Batwa-ecology swelled at that moment to something approaching awe. With diligence and a lifetime to practice, I'm sure I could learn to track Katendegyere group through a dry season forest. But I'm equally sure it would take me more incarnations than a Hindu folk tale to master the art of tracking bees.

An hour later I had given up on finding a clear view of the gorillas. They were aware of our presence and seemed unconcerned, and that was all we could hope for in the thick bush. I leaned back against a tree stump and closed my eyes. The trackers were discussing something in whispers when we heard a sudden grunt and the noise of a gorilla moving our way with purpose. Without looking we knew that it had to be Makale. Scrambling to our feet we watched him approach like a wave parting the greenery. He hadn't charged in ten days and we all crouched submissively, hoping to extend that streak.

Shoving down a wall of leaves and branches he came suddenly into view and stopped, appraising us with a dark-eyed glower. His pungent horse-and-sweat reek surrounded us, but he didn't charge. We backed away slowly to our fifteen-foot distance and Makale seemed satisfied. With a hostile glare he pushed the vegetation aside and stalked through the open place before us, stiff-legged, like a gunfighter at the swinging doors of an old west saloon, all chest and swagger and heavy brow.

Similar to charging or the chest-beat sequence, the 'stiff-legged walk' displays a predictable level of gorilla aggression. Males reinforce the dominance hierarchy within the group by using these visual signals in minor conflicts over feeding privileges or right of way. Subordinate animals will either give way or return the gesture to escalate the challenge. The scale of the display indicates the animal's level of anxiety, and used alone, stiff-leg walking is a relatively minor signal. Makale was annoyed with us, but not particularly angry. He moved directly to an open area nearby and began feeding.

We followed at a distance. Feeding in the midst of conflict can be a type of 'displacement behavior,' where the individual's aggression is redirected temporarily into a non-threatening activity. Maintaining our presence with Makale was essential to advance his habituation, but none of us wanted to push him into charging. He kept one eye on our cautious approach, then turned away from us, crossed his arms over his chest and passed gas, a long, disdainful movement of air.

With that gesture he dismissed us completely and focused on his meal. Soon others came into the clearing: Karema, Mutesi, and a young blackback I recognized only from the parallel scars marking his nostrils. They were feeding on green vines that climbed and tangled in the low, shrubby branches. Tearing down handfuls, they pulled each strand through their wide mouths, neatly stripping off the leaves with an audible snapping sound.

The second hour of habituation went perfectly, with the gorillas feeding, resting and interacting among themselves as if we weren't even there. We saw Mutesi displace Karema from her feeding site with a belligerent lunge and series of pig-like grunts. She moved a few meters off, dragging a fistful of vines behind her, and continued feeding, apparently unperturbed. Makale watched the exchange with indifference before rolling onto his back to rest, one leg splayed upwards like another dark trunk in the shadows. The younger blackback remained slightly apart from the others and moved out of sight after a few minutes. He was still maturing and spent most of his time near Mugurusi, the shy lead silverback. We rarely saw either of them clearly, and the trackers knew him only as 'the small Makale.' He and Makale resembled one another closely and were probably brothers, but I hoped that the likeness only ran skin deep. With two Makales in the group, we'd all be professional bee trackers before Katendegyere was ready for tourism.

Rain fell in heavy droplets and the gorillas had retreated into a thicket when we finally backed away and started the long hike home. Prunari and Charles smiled and joked the whole way in spite of the weather. Every good day of habituation brought us closer to opening the group for tourism on schedule, an event that would bring both of them healthy bonuses from the gorilla program (IGCP).

We paused for lunch under the dry eaves of a huge rainforest tree, its buttressed roots curving down around us like the arms of a starfish. Between mouthfuls of cold *matoke* and beans, Prunari gestured towards the forest and mimicked a gorilla scream, reenacting stories of habituation past. Charles laughed and joined the pantomime, slapping at the ground with two hands and tracing paths through the foliage where the gorillas had once charged in this very spot. The entire group had encircled them, chest-beating, tearing up vegetation and charging repeatedly while the trackers cowered by the tree for over an hour.

"But not now," Charles concluded. The incident had taken place nearly two years ago, after only a few months of habituation. "They can't charge now," he repeated, "they are too calm."

Several weeks later we announced the beginning of our 'experimental tourist' phase. The gorillas had been consistently passive, and we wanted to gauge their response to a series of new faces. By taking in two visitors per day, we hoped to show that the group could safely open for general tourism at the end of the month. If the influx of new people produced notable signs of aggression or fear, we could simply stop the experiment and continue with our regular habituation.

We had no trouble finding guinea pig tourists. Complimentary tracking permits brought local community leaders and government officials, as well as a host of expatriates working for the national park system. Eager to view mountain gorillas in the wild, they all ignored the stipulation that Katendegyere group was still under habituation, and that tracking could be cancelled at any time depending on their behavior.

One of our first visitors was a good friend of mine from Kampala. Ted Hazard worked as a financial advisor at park headquarters, and his two-year contract nearly overlapped my own. I'd stayed with him several times in the city, where good humor, hot showers and a toaster made his house a regular Peace Corps motel. With his brother visiting from San Diego, Ted had organized a trip through several parks around the country, culminating with a visit to Bwindi and a chance to track Katendegyere group.

"You see, these gorillas are only partially habituated," Medad warned them, "and sometimes they can change their minds." He was halfway through the tourist briefing, explaining in careful English the strict procedures for gorilla tracking in Bwindi Park.

"Oh, great." Ted exchanged a look with his brother and they laughed, with only a hint of worry tracing their sarcasm.

"If the gorillas charge, please you don't cry out or try to run. No. Just crouch down and wait. Follow our example and when the animal is calm, we will go slowly, slowly backward to our distance of five meters."

The gorillas had spent several days in a steep, shrub-choked valley outside the forest and we hiked for half an hour through morning sunlight and broad-leaved banana plantations. Worn smooth by generations of sandals and bare feet, the path wound across the hard red earth, connecting fields, families and farms. Collared sunbirds darted through the air above our heads, searching for dew and nectar in the drooping clusters of banana blossoms. The trackers diverted from the main pathway and we began following a dry creekbed, ascending slowly towards a fragment of rainforest at the head of the narrow valley.

The rasp of a saw echoed rhythmically through the air and we soon met a man carrying freshly-hewn mahogany planks. He balanced the load easily on his head, nimbly crossing the rocks in bare feet as he descended towards the village. Medad and the trackers greeted him and exchanged a few words. The timber was being cut to order for several carpenters in town, where the demand for hardwoods far outstripped the supply. When Bwindi was still a forest reserve, loose regulations and large scale corruption ensured a steady flow of illegal timber to lumber merchants throughout the district. Mature hardwoods disappeared from much of the forest before the park was created, and most stands of native woods outside the reserve were cleared altogether. The few mature trees remaining in this valley would soon be gone, cut and sold by their owner, an irascible old farmer named Behuari.

My house at the edge of the forest

Makale

The Impenetrable Forest

Walking home through millet fields

Family shambas in the hills around Bwindi

Mutesi

Kasigazi

Kacupira feeding on banana stems

The park staff

Kashundwe and Bob, mother and son in Mubare group

Mugurusi, lead silverback of Katendegyere group

Makale

Karen Archabald, the author, Tibesigwa Gongo and John Dubois

Karema

Makale meets a tourist

Pit sawyers at work in Behuari's valley

First aid training with ranger-guides Enos Komunda, Betunga William, Alfred
Twinomujuni, Hope Nsiime and Levi Rwahamuhanda (from left)

Tracker Prunari Rukundema counts banana stems destroyed by Katendegyere group

Kacupira

The author and Phenny Gongo

Kasigazi at play

A party in Buhoma: Prunari, Medad, Abel, Stephen, George, Philman, Phenny, Caleb, Levi, Alfred and James (from left)

Neighbors and friends on the Ntale family lawn: Millie, the author, Rob Rothe,
Emma, Susan, Sam, Tom and Aunt Florence (from left)

Less than twenty years ago, Behuari's valley was connected to Bwindi by a large forest that stretched for several miles along the ridgetop west of Buhoma. The area formed a central part of Katendegyere group's home range, and although most of the trees have fallen prey to the demand for wood and seasonal croplands, the group still crosses large distances of open space to visit patches of forest like Behuari's. More than once we sat with the gorillas and heard a thundering crash of timber as another piece of their dwindling rainforest home was literally cut down around them. The situation seemed almost contrived, like an elaborate fundraising advertisement for Greenpeace or the World Wildlife Fund. It was, however, unfortunately and poignantly real, a sad microcosm of conservation problems in Uganda, where the rapidly expanding human population is pressing right up to the edge of protected areas.

IGCP and the park had both offered far above market value for Behuari's land, but he refused to sell. He owned other *shambas* closer to the village, but this valley was an important family asset, something more tangible and lasting than any one-time payment of cash. In desperation we offered him a logging contract, agreeing to purchase all the timber in the valley with only one stipulation, that he left it standing. This too failed and the trees continued to fall.

As we climbed up the valley, the banana fields thinned to a few ill-tended clumps clinging to the steep hillside. The guides had told me that Behuari planted here with no expectations of harvest. He knew the gorillas often raided these outlying patches, and wanted to supplement his logging income with crop damage compensation from the park. With Behuari, we had learned that the pronunciation of his name, "be wary," was good advice for any business negotiations.

The trail narrowed and turned up the northern side of the valley and we came across a group of banana trees where the gorillas had been feeding. Behuari himself stood waiting for us, wiry and fit in his ripped blue T-shirt and rounded knit cap. A scraggle of grey beard clung to his narrow chin and he had the calloused, cracked hands of a long life in the fields. He blocked the trail as we approached, arms akimbo, and gestured at his ruined bananas with an expression of stern horror.

The trees lay scattered about in piles of ropy white strands as if exploded from within. While Behuari lamented his loss to the trackers, I stooped and tugged up a handful of the crushed stem—wet and fibrous, like coarse, overgrown cornsilk.

The old farmer continued his indignant rant until I urged Medad to move ahead. "Tell him we'll come back to count the stems later," I said. The park offered a small sum to compensate farmers for every lost tree, and encourage them not to harass the gorillas on their land. Medad translated and Behuari nodded his grudging approval. For fun I added, "Tell him he shouldn't plant his bananas in the gorillas' *shamba* ."

He looked at me warily for a moment, then barked a laugh at my crazy *muzungu* joke. I smiled too and watched him disappear down the trail, finding myself liking the man in spite of his determination to clear the valley.

Just then we heard another kind of bark, the unmistakably resonant cough of a silverback far above us on the valley wall. Climbing towards the noise, we crossed the charred lunar landscape of a newly burned millet field. Every footstep kicked up plumes of grey ash and knocked a small slide of stones and dirt down the steep slope. Much of the upper reaches of the valley had been planted at one time, and lay now under a thick tangle of weeds, grass and shrubs. Millet is hard on the soil and fields must lie fallow for up to four years before they can support another crop. To transform the pillared greens of the rainforest into this patchwork of bracken and soot seemed a poor exchange indeed.

We paused to catch our breath on the heat-baked, near-vertical grade, and it struck me as a particularly terrible place for a farm. Mere walking challenged my sense of balance, let alone swinging a hoe or harvesting heavy bushels of grain. People didn't plant on the steep hills around Bwindi because it was good land, or because they had a special desire to ruin gorilla habitat. Only one reason drove them to farm such rugged terrain: in southwest Uganda, there was little place else left to go.

The trackers beckoned to us from above. They'd found a fresh trail and we advanced together slowly, tunneling into the thicket. The gorillas were nearby, but we could see only a few feet through the dense undergrowth. I walked in the lead with Mishana, peering into the vegetation, disappointed that we might not get a clear view for Ted. Finally we saw an indistinct flutter of leaves and heard the long sigh of a gorilla at rest. Circling to approach the sleeping animal from above, we could just discern its dark form lying quietly in the shady gloom.

Suddenly the greenery erupted with screams and a surge of animate shadow as Makale charged full speed towards us from somewhere up the hillside. We crouched in a tight group and he loomed over us with a deafening roar, his eyes wild and angry and his teeth bared, two rows of sharp white jags in the huge red of his mouth. After an endless, gut-wrenching instant he settled back on his haunches several feet before us, squatting and glaring like some dark vision of a belligerent Buddha.

His nervous stench filled the air with a thick, horse-and-sweat perfume as we waited for the tension to pass, holding our ground, but looking downward submissively. Makale followed our gaze to the ground and noticed the *panga* that James had dropped during the initial charge. With an almost contemptuous grunt he leaned forward and snatched up the blade.

From the standpoint of developing safe, ecologically responsible tourism, this was a low point. Our first visitor from park headquarters had been instantly charged, and now the gorilla was armed. I turned to Ted and attempted an encouraging 'no problem' smile while Makale twirled the long knife in his hands, alternately sniffing and biting the handle and blade. Several minutes passed and I had visions of him cutting himself on the sharp steel, but he handled the *panga* gently, even nimbly in his huge, leathery fingers. Abruptly contented with his inspection, Makale threw the blade back to the ground at our feet, his expression seemingly tinged with a look of smug satisfaction.

Behind me, Medad was sitting on his walking stick, trying to keep it hidden under a layer of trampled vegetation. Makale picked him out immediately and reared up, screaming, to lunge over our heads and grab at the stick. Medad didn't hesitate in taking the only logical course of action: he let him have it.

Makale gripped his new prize and moved a few feet off, sniffing and biting the smooth wood as he had the *panga*. He held it out in front of him and peered along its length as if inspecting a rifle barrel for flaws. I cringed every time the stick passed into his mouth. In a project designed to minimize the risk of disease transmission, gorillas licking recently-handled equipment ranks high on the list of catastrophes. When Makale moved towards us again we held our collective breath, and I tucked my camera under my jacket, dreading what new object might catch his attention. But he came only to return what he'd borrowed, leaning forward and gingerly sliding the walking stick back into place under Medad, exactly where he'd found it.

With a final haughty glare, Makale left us, striding nonchalantly up the slope. His broad back mingled quickly into deeper shadow, and there was a moment of silence in the thicket, like the heartbeat pause between a good punch line and laughter. Nervous energy drained from the air in a rush and we found ourselves suddenly smiling and shaking our heads with relief and awe.

Below us, the first gorilla turned over with a soft rustle of branches and slept on, completely oblivious to our encounter with its comrade. The whole drama had taken less than twenty minutes and we could have pressed on, but Ted and his brother looked like they'd seen enough, and none of us were too excited about following Makale deeper into the undergrowth.

"Well, that's what we mean by partially habituated," I whispered as we backed slowly away.

"So," Ted replied, never without a comeback, "when they're fully habituated, do they pick up the machete by the right end?"

"Exactly."

Actually, Makale's habituation was making definite progress. We had obviously progressed beyond the 'fear' stage. Our presence still annoyed him on occasion, but he no longer found us particularly alarming. The new challenge would be to control his boundless curiosity. Luckily, equipment inspections didn't become a habit, and the rest of the month passed without incident. Katendegyere group opened for business on time, the trackers got their bonus, and our guinea-pig tourists were replaced by paying customers. We concluded the 'experimental tourism' phase with two significant results: a new regulation requiring people to leave their *pangas* and walking sticks at least 200 meters away from Katendegyere group, and a lingering rumor that "armed gorillas" were roaming the depths of Bwindi Forest.

CHAPTER X

A STRANGE MAN

―――――▼―――――

*"You know what I always was—made up of queer materials,
and averse to beaten paths; unfortunately, not fitted
for those harnessed positions which produce wealth; yet,
ever unhappy when unemployed, and too proud to serve..."*
—Samuel Baker, before leaving to explore the Nile, *1861*

On a fine morning in the dry season, I set out with the trail crew to build foot-bridges over the Muzabajiro River. Clear weather allowed for a full day's work and we always scheduled major construction projects for the dry months. The Muzabajiro wound its way three times through our new loop trail, and even the trackers had slipped crossing its slick, smooth stones. Eventually, the Muzabajiro Loop would become a major path for gorilla tracking and forest walks, but we needed sturdy stream crossings before we could open the route to tourists.

Bwindi's first trail crew varied more than Makale's mood swings or the price of eggs in Kihihi. From young men paying off dowries or drinking debts to teenagers earning their school fees, a crowd gathered at the office each morning for the chance at a day's wage in the forest. The chaos of picking out a crew involved long heated arguments and a continuing lesson in village politics. Finally, at the risk of playing favorites, I began choosing from a small group of regulars, hoping to one day hire a permanent team.

Climbing now through the lush green tea fields below Dominico's farm, I paused to look back at the day's crew. Each man carried a heavy building pole and a sharp, double-edged *panga*, the universal tool of rural Africa. They hauled the poles lengthwise on their heads, balanced on tiny cushions of woven banana leaves. I had tried this only once, and gave up after a brief stagger with screaming neck pains and one firm conclusion: hiking the winding trails of Bwindi with a tree on my head was a skill I'd never master.

I greeted the men as they passed me on the narrow track, relying on one of the only practical vestiges of colonialism, English names. Most people in Uganda took a Western name in addition to their family and tribal titles. The combinations often sounded incongruous, but when the local alternative was a tongue-tangled attempt at Rwahamuhanda or Kawermerwa, it was a huge relief to be able to call someone Stan.

Today, Noah was the crew's nominal leader. He had finished primary school and knew a bit of English, which, combined with my Speak & Spell Rukiga, was a passable form of communication. Richard and Samuel followed him up the trail, two rangy, slightly nervous-looking youths in matching sky-blue sweaters. They claimed to be students, although I never saw them any closer to a schoolhouse than the local *tonto* bar on market day.

Stanley brought up the rear and gave me a big smile from under his head-load. "*Agandi*, Boss!" He was about my age, hard-working, and commonly regarded as the strongest man in Buhoma. "Boss" constituted the bulk of his English vocabulary. He used it as a sort of universal noun and managed to fit it into nearly every sentence we exchanged. As far as I could tell, I was 'Boss,' he was 'Boss,' the gorillas were 'Boss,' and so was his *panga*. I often wished that I knew a similarly versatile word in Rukiga.

From Dominico's place the trail wound through abandoned millet fields in a steep valley along the creek. The forest rose on one side, towering and ancient in its shade, while lush shrubs and saplings erupted into life in the clearing. We walked through a sea of neck-high greenery, pale and new, and scattered across with white blossoms like handfuls of tiny pearls

As we descended toward the forest edge I mused about entering an African jungle with companions called Noah, Richard, Samuel and Stanley. Quite an illustrious group in the annals of adventure, from the greatest biblical wanderer to three of the most intrepid European explorers to ever reach the African continent: Richard Francis Burton, Samuel Baker, and Henry Morton Stanley.

To the outside world, nineteenth century Africa truly was The Dark Continent, a blank enigma on every map. Port cities like Zanzibar and Mombasa gained fame as exotic centers for the slave and ivory trade, but the interior remained a mystery. Merchants relied heavily on middlemen among the coastal tribes, and only penetrated inland when the nearby stocks of elephant and slaves were exhausted. These traders, mostly Arabs and Turks,* pioneered exploration in East Africa, establishing caravan routes deep into the Sudan and across modern day Tanzania in the early 1800s. While they held commercial goals above cartography, tales of great mountains and vast inland lakes accompanied their shipments of ivory to the coast. When these stories reached Europe, adventurers began preparing to chase the ultimate geographical prize of the century: discovering the source of the river Nile.

On setting out from Cairo in 1861, Samuel Baker wrote: "Before me—untrodden Africa; against me—the obstacles that had defeated the world since its creation." For millennia, the Nile's headwaters eluded scholars from around the civilized world. Ptolemy, the great historian of ancient Greece, drew a map in the second century A.D. that put the source in a theoretical range of high peaks he called 'The Mountains of the Moon.'

---

*   Turkish caravans consisted largely of Egyptians, who were ruled by an Ottoman Khedive until 1914. They followed the Nile south into the Sudan and northern Uganda, while Arabs generally forged inland from the East African Coast.

Since that time, no one had been able to confirm or refute his claim, all explorations turned back by hostile tribes and an unforgiving landscape. To people in Europe and America, venturing into the African interior was an almost unthinkable feat, the equivalent of modern space travel. And like astronauts, the early explorers crossed into the unknown driven by a fundamental curiosity, a fascination with the natural world, and a not-insignificant desire for personal glory.

Noah aside, the namesakes of Bwindi's trail crew all played significant roles in the quest for Nile. Burton paved the way, travelling by foot from the coast to Lake Tanganyika with his partner, John Hanning Speke in 1858. Their two-year journey together was inconclusive, with Burton touting Lake Tanganyika and Speke claiming Lake Victoria, his own discovery, as the river's true source. The argument dissolved their partnership and Speke later returned alone to successfully follow the Nile from Lake Victoria north to Cairo. He spent several months in modern-day Uganda along the way, and eventually met the Bakers, Samuel and his indomitable wife Florence, who had traveled up the river as far as a Turkish trading post in Sudan.

The Bakers proceeded southward to support Speke's findings, and navigate the length of Lake Albert, a secondary source for the river. Still, controversy raged for over twenty years, with Burton, David Livingstone and other scholars doubting Speke's geography, and forwarding their own headwater theories.

Only the journeys of H. M. Stanley finally put the matter to rest. Called *Bula Matari*, "The Rock Breaker," by his native guides, Stanley's legendary stamina led him across the continent twice between 1874 and 1889. While best remembered for locating David Livingstone deep in the interior with the immortal line "Dr. Livingstone, I presume?"—Stanley also settled the Nile issue. Setting out from the East African coast near Zanzibar, his 1874 expedition brought a sturdy wooden boat, carried in sections. He used it to circumnavigate both Lakes Victoria and Tanganyika, proving once and for all that Speke's discovery, Victoria, did indeed run north to the Nile. Tanganyika, he learned, flowed westward, and he followed its muddy drainage all the way to the Congo River and on to the Atlantic.

None of these explorers ever reached Bwindi forest itself, but Baker and Stanley both crossed into Uganda, and Speke passed within sight of the "bold sky-scraping cones" of the nearby Virunga Volcanos. Geographical notes and the trials of the trail filled up much of their journals, but all the explorers found time to wax lyrical about the landscapes and people they encountered. Of trudging through uncharted rainforest several hundred miles west and north of Bwindi, Stanley wrote the following:

> *"Imagine the whole of France and the Iberian Peninsula closely packed with trees…whose crowns of foliage interlace and prevent any view of sky and sun… Then, from tree to tree run cables from two inches to fifteen inches thick, up and down, in loops and festoons, and W's and badly formed M's; fold them round the trees in great tight coils…like endless anacondas; let them flower and leaf luxuriantly and mix up above with the foliage of the trees to hide the sun;…and at every fork and at every horizontal branch bend cabbage-like lichens of the largest kind and broad spear-leafed plants…and orchids and clusters of vegetable marvels, and a drapery of delicate ferns which abound. Now cover tree, branch, twig and creeper with a thick moss like a green fur."*

Passing now under the green eaves of Bwindi's great canopy, the temperature dropped a welcome ten degrees and the world dimmed to a realm of shadow and sunflecks. In spite of dry weather, the air felt moist and smelled of rich, damp earth. Cricket song and the hammernotes of a tinkerbird punctuated the creek's white rushing, and for a moment the scene seemed pristine, like a page from Stanley's journal or sometime before.

In Africa, one can easily succumb to landscape nostalgia, a deep longing for the past. The vast wilderness mapped by early explorers represented more than just an unspoiled ecosystem. To naturalists it was a glimpse into distant history, an almost Pleistocene world where humanity still played a minor role; living in small communities; hunting, but also hunted. The population explosion in this century has shrunk such areas to pockets like Bwindi, surrounded by cultivation and threatened with human pressures. The constant struggle to preserve these remnants makes historical wilderness an all the more appealing fantasy.

Certainly, the passage of time distils complex events and simplifies our perceptions of history. The visceral threats of just being alive give each generation its own sense of critical imperative, and we live convinced that problems were far simpler in 'the olden days.' That view may be distorted, but the rapid environmental decline of recent decades is truly an unparalleled event in human history. Wild places are disappearing around the world, taking thousands of unique plants and animals with them each year. Wilderness itself may become a historical concept in the next century, and the long term prospects of rare species like mountain gorillas will always be in question.

Working in the shadow of such thoughts, I often took comfort in picturing a Bwindi primeval, and today, accompanied in name by four great naturalist-explorers, I imagined the rainforest stretching uninterrupted around me, north along the rift escarpment, south to the Virunga volcanos and west across the Congo basin to the sea.

"Noah," I called out, "how do you choose these *muzungu* names?"

He looked at me questioningly, somehow managing to raise his eyebrows under the heavy eucalyptus pole.

"The Christian names," I said, searching for a clearer Ugandan phrasing, "how do you pick them?"

"Ah," he said and dropped the pole to the ground with an elaborate shrug. We had arrived at the first stream crossing. "Me? From the Bible."

"Right," I nodded in return. This was no surprise. Protestant and Catholic missionaries had traveled far and wide in Uganda, and still operated numerous schools, hospitals and orphanages. Over 85 percent of the population follow Christian beliefs, and a burgeoning evangelical movement adds more converts to the flock every year.

The others had arrived and Noah pointed to each of them in turn as they dropped their burdens.

"Burton, Stanley, Baker," he identified in rapid succession.

I was floored. "You know these explorers?" I asked, "these Europeans?"

"Yes, yes," he smiled enthusiastically. "Big African whites. We learn them in grade three. And Emin Pasha, Mr. John Hanning Speke…"

He went on to name several explorers I'd never even heard of, men who, by the vagaries of a colonially-designed education system, are still remembered, even honored, in the remote areas they traversed over a century ago. Their legacy seems all the more ironic in light of the pages they devoted to disparaging Africa's varied 'savages.' Consider Baker's summary of the Sudanese: "…the most lying, perfidious, mean, dirty scoundrels on God's earth. They are all alike, therefore it is no use kicking the posterior of an individual." Or Burton's high opinion of the Wafanya: "They were pertinacious as flies; to drive them away was only to invite return, while— worst of all—the women were plain, and their grotesque salutations resembled the 'encounter of two dog-apes.'"

The locals, in turn, weren't always impressed by their European visitors. Early explorers should be glad to be known for something in addition to inspiring the name '*Muzungu*,' a lingering title applied to all light-skinned people in East Africa. The word comes from a Swahili verb meaning 'to move around in circles,' referring both to the Westerners' frantic pace in life, and their propensity for getting incredibly lost.

I unloaded hardware from my backpack and we started to work on the bridge. The Peace Corps and other international development agencies stress the need for 'appropriate technology,' using designs easily replicated and maintained with local materials. Too many projects have failed due to a lack of tractor parts in New Guinea, or the high price of unleaded gas in rural Mozambique.

In this regard I considered myself particularly well-suited to development work. After a cursory lesson, any trail crew member could replicate
and probably improve upon my simple footbridge designs. The machinery
involved—nails, hammer, *panga*—was necessarily 'appropriate' for one
simple reason: I was incapable of operating anything more complicated.

Two large timbers already spanned the stream from our work the previous day, and we spent an hour gathering stones from the creekbed to help
support the muddy bank. Then Richard and Samuel started cutting and
splitting poles while Noah and I hammered them onto the emerging
bridge. Stanley continued gathering rocks, using his uncanny strength to
wedge torso-sized boulders under the supports.

It was at this point, suspended above a rushing rainforest stream, hammer in hand, that it hit me: *I'm in the Peace Corps, building bridges in
Africa. I could be the poster child for this outfit.*

Tom Hanks and John Candy once starred in a Peace Corps parody
called "Volunteers." Hanks played Laurence Bourne III, a cynical socialite
who joined the Peace Corps to escape his gambling debts. He summarized
his feelings for the experience in four words: "So, this is Hell."

Their task was, of course, to build a bridge across a river in some
steamy equatorial jungle. I remember at the time identifying strongly with
the Bourne character, thinking that, yes indeed, life couldn't get much
worse than that. Ten years later I found myself on the Muzabajiro River,
hammering away with the Bwindi Forest trail crew and thinking that life
couldn't be much better. I don't know, maybe the heat was getting to me.

After several hours I left the trail crew hard at work, confident in their
knowledge of bridge design, and their ability to bend far fewer nails than
myself. I hiked ahead into the forest to scout a new section of the loop
trail. Most of the route had already been chosen and cut, a patchwork of
new paths and old poacher trails winding up the slopes of Rukubira hill.
The forest was still and somnolent in the heat of midday and my steps
rasped through a layer of dry leaves. The noise startled a tambourine dove

into wing-snapping flight, and I watched its pale feathers wheel and vanish quickly into the dense shade.

The harsh sound of my passage startled me as well. A typical footstep in Bwindi landed softly in mud and a sodden compost of dead vegetation, but weeks without heavy rain had left the ground cover as crisp as autumn leaves. Hundreds of forested acres burned during just such a dry spell in 1992, when farmers lost control of a blazing field bordering the park. The fire devastated vital habitat for most rainforest creatures, but luckily created an excellent feeding ground for gorillas. Purple-flowered *Ipomea* vines soon dominated the clearing, creeping over shrubs and twining up the silvery dead tree trunks. Both Mubare and Katendegyere groups frequented the area to feast on the lush green leaves.

My path came suddenly into the open, a hilltop arm of that same burn scar, and the forest stretched away southeast before me: deep green ridges hunching against one another into the distance. The furthest hill lay outside the park, I knew, and was cleared long ago for agriculture. But through the haze its cultivated greens were indistinguishable from rainforest, and for a moment I could almost believe my fantasy of the Bwindi primeval.

Clouds gathered in high pillars above the hills, and the afternoon's first thunder rumbled through the air like a summons. Even in the dry season, lightning storms were common in Bwindi and people regarded them with seeming indifference. But I learned that their nonchalance was tinged with respect, and a well-deserved touch of fear.

I once returned from a trip to Kenya with several dozen hammered copper bracelets, thinking they'd make good Christmas gifts for the park staff. The trackers and guides received them with great enthusiasm, thanking me profusely as they clasped the thin strands around their wrists and held them up in the sunlight.

Soon after, I noticed a dramatic increase in visitors to my house. Suddenly everyone in Buhoma was very interested in my return, stopping by to welcome me back and wondering if, by the way, maybe I'd brought them something from Kenya? Unfortunately, I hadn't been shopping with the whole county in mind, and had to turn people away empty-handed. I knew my popularity was plummeting in the village polls.

Dominico, in particular, was adamant. He stayed for over an hour, berating me with high-speed Rukiga exclamations. He would stare meaningfully up at the sky, then rub his barren wrist and frown, the downward furrows of his old face writing their own volumes of reproach. He was not mollified by a cup of milk tea. He did not want a notepad. I was relieved, finally, to see Alfred Twinomujuni, one of the guides and a former school teacher, coming up the path.

He listened to Dominico for a moment, then turned away and began flipping idly through a magazine on my desk.

"This one is only superstition," he told me with some disdain. "They say the bracelets will protect against lightning."

"Lightning?" I stared at him blankly. That explained Dominico's wrist and sky gestures, but I wanted more. I wanted it to make sense. "How can a bracelet repel electricity?"

"I don't know," Alfred shrugged. "It's a tradition."

Ephraim arrived at that point, laughing under his hand as the old man launched into another long, beseeching rant.

"He says you want him to be killed."

"What?!" I tried not to sound too exasperated.

"By not giving the bracelet. He is now sure to be hit by the lightning," Ephraim translated. Dominico nodded significantly, imploring me with another frown.

To those of us raised in temperate climes, being struck by lightning is hardly a pressing concern. People worry about street crime. They worry about cholesterol. When a few small pieces of the Skylab space station started raining down from the stratosphere in the late 1970s, people worried. But lightning strikes rank in the league of alien abduction or spontaneous combustion as a truly unusual way to go.

In Bwindi, I learned, this is not the case. Thunderstorms are a daily occurrence, like the sky breathing, and lightning strikes really do take lives. John witnessed the aftermath of one such strike while walking to Butagota. A crowd had gathered around the bodies of a man and child under the still-smoking tree where they'd taken shelter from the rain. For months afterwards John ran for cover at any sign of a storm, and we had lively debates on the relative safety of banana thatch or metal-roofed houses. Local people are haunted by such events, and any precaution, even an obscure bracelet story, was worth the effort.

I had only one bracelet remaining, and as I hooked it around Dominico's wrist, his face lit up. "My son! You!" he cried and walked out into the sunlight, thrusting his copper band up at the sky like an aegis. "Good!!"

When he was gone I turned to Alfred. "So now the rest of the village thinks I want them dead?"

"You just don't mind," he told me. "These are superstitions."

We chatted about other things for a while and he borrowed a magazine, but before leaving he paused by the door and wondered aloud: "Maybe you have another one of the wire bracelets?"

I glared at him and he smiled sheepishly, adding, "for my wife."

Late afternoon sunlight angled in rich golden beams through the forest when I completed the loop and returned to check on the trail crew. I found Noah and Richard hammering the last struts onto a fine looking bridge.

"*Mwakora!*" I said in greeting, a Rukiga phrase to thank someone for working.

We walked across the span together and it was sturdy under our combined weight.

"Very nice," I told them. "This one is strong."

On the opposite shore, the trail curved out of sight through a thick stand of *Mimsops*, a tangling, neck-high bracken. I heard the others laughing and talking somewhere ahead, and then, oddly, the sound of a hammer. I walked quickly onward and stopped in disbelief ten yards up the trail at another new bridge, nearly finished, leading directly back across the river.

Stanley and Samuel were standing knee deep in the rushing stream, hammering and talking loudly, but they paused, suddenly wary when they saw the look of shock on my face.

"Noah," I began, but faded off in frustration. I was torn between amazement at how much they had accomplished, and dismay at wasting so much effort on a bridge to nowhere.

"Yes," he replied with a broad smile.

"Noah, this is a very fine bridge," I lied, frowning, but trying to be positive.

"Yes," he said again, more hesitantly as my agitation became apparent. The others gathered around us with expressions of helpful concern.

"A very good bridge," I began again, "but where is it going?"

"What?" Noah looked genuinely worried now.

"Noah, there's no trail here," I finally snapped. "Where is this bridge going?!?"

He paused and looked thoughtfully at the bridge in question before turning back to me and replying gravely, "I don't know. I am a very strange man."

"Ah," I smiled inanely, waiting to understand his comment. Noah smiled too, and the others nodded their general assent. Slowly, as Noah's words continued to make no sense to me whatsoever, I realized that our conversation was probably over. Strange man. Where do you go from there?

"Great work!" I finally said, "Yes!" and gave them all a big thumbs up.

What else could I do? We re-routed the trail to include the superfluous bridge and I notched another communication triumph into my cross-cultural belt. Laurence Bourne III was laughing cynically in my head. Building bridges to nowhere; now I was a *true* Peace Corps Poster child.

CHAPTER XI

BAUMGARTEL'S CAR

▼

*"I thought of the lamp as a friendly genie, particularly when
stepping outside the tent into a bitterly cold, ink-black night. It
was awesome to think of this as the only speck of light, other
than perhaps occasional poacher fires, within the entire Virunga
mountain range. When contemplating the vast expanse of
uninhabited, rugged, mountainous land surrounding me
and such a wealth of wilderness for my backyard,
I considered myself one of the world's most fortunate people."*

—Dian Fossey
Gorillas in the Mist, 1983

Translated loosely, the word gorilla means "scratcher" in a long-dead language spoken by the people of ancient Carthage. When their greatest
explorer, Hanno, led a fleet of ships down the West African coast in the
fifth century B.C., one of his shore parties encountered a troop of strange,
man-like beasts in the forest. He described a battle where several of his soldiers were attacked, and badly 'scratched' trying to capture the unknown
animals. They succeeded in bringing two skins back to Carthage, and
although the apes were probably chimpanzees, or even some type of monkey, the name has been associated with gorillas ever since.

In West and Central Africa, forest people have coexisted with gorillas for countless generations, revering them as clan totems, or hunting the apes as a prized source of food and powerful fetish medicines. To the outside world, however, the gorilla remained a mystery for more than 2,000 years after Hanno's initial discovery. Stories filtered north to Europe through trade routes and early explorations, but even with the increase in African travel brought on by slavery, Westerners didn't officially recognize the species until 1847, when a pair of Protestant missionaries in Gabon gathered the first skull samples and sent them back to England. Mountain gorillas lived in even greater obscurity, over a thousand miles east of their lowland cousins' known range. They eluded scientists for another five decades, until an insignificant military expedition visited the Virunga Volcanos at the turn of the century.

In 1902, several years after Germany laid claim to modern-day Rwanda and Burundi as part of its East African territory, Captain Oscar von Beringe led a small contingent of troops into his country's newest protectorate. Meant as a show of force to intimidate the local rulers and impress border guards of the neighboring Belgian Congo, Beringe's mission would make its most far-reaching impact in the realm of science. While encamped on the slopes of Mt. Sabinyo, he shot at a group of black apes, and managed to bring one specimen back to Germany. The skeleton created an immediate stir, confirming a long-standing rumor that large, thick-haired gorillas were living in the highlands of Central Africa. Taxonomists declared a new subspecies, the mountain gorilla, and honored Beringe as its discoverer. His journey's political goals came to nothing when Germany lost 'Ruanda-Urundi' to the Belgians in World War One, but Beringe's name lives on in African history, preserved in the scientific title of his simian quarry, *Gorilla gorilla berengei*.

In the decades that followed, museums and zoos from around the world strove to add this rare new ape to their collections. Naturalists and sportsmen began mounting expeditions, and each year brought more safaris to the once-obscure Virunga Volcanos. Most travellers entered the region through Uganda, where passable roads stretched inland as far as the British colonial outpost at Kabale, only a three or four day trek from the mountains. This route passed southeast of Bwindi, through a landscape of steep, rounded hillsides and tiny villages that European visitors soon dubbed, "the Switzerland of Africa." With the volcanos looming over every turn in the path, and small eruptions from Nyamulagira lighting up the sky at night, the countryside gained quick fame for its almost-mythical natural beauty. American naturalist Mary Akely described her 1926 journey with reverence:

> "At sunset we had a wide and glorious view into deep, green valleys and blue lakes–beautiful Mugisha, Tshahafi and Bulera. We looked across a vista of hills under cultivation by the natives and over bamboo forests to the gilded pinnacles of the extinct volcanoes of Sebyinyo, Mugabura, and Mugahinga,* standing as outposts on the Uganda-Congo frontier. Again and again throughout the moonless night the lurid beacon fires of active Nyamlagira flared and fell, while the vast silences were broken by the faint and doleful throbbing of the drums. Surely this was not a night meant for sleep!"

Akely's husband Carl, who died in the Virungas on that same expedition, is remembered as a pioneer in the world of gorilla conservation. His specimens for the American Museum of Natural History portrayed the apes in authentic poses, offering a realistic alternative to the violent image of gorillas popularized in folklore, travel literature, and the top-grossing film of 1933, "King Kong." After his first collecting trip in 1921, Akely helped convince the Belgian government to include their Congo colony's

---

* Various spellings for the volcanos have appeared over the years, but most modern maps list the six main peaks as Muhavura, Mgahinga, Sabinyo, Visoke, Karisimbi and Mikeno, with Nyamulagira and Nyiragongo rising up 20 miles to the west. Also, some cartographers still label the entire Virunga chain as the Bufumbira or Mfumbira mountains.

portion of the Virunga Volcanos as a permanent gorilla sanctuary in Parc National Albert, the first national park on the African continent.

Today, the Virungas lie within protected areas in three countries: Parc des Virungas in Zaire, Parc des Volcans in Rwanda (both formerly part of Parc National Albert under the Belgians), and Uganda's newly-created Mgahinga Gorilla National Park. The footpath from Kabale has been replaced by a ribbon of red-dirt road, but travelling to the Virungas is still a trip of staggering beauty. When Mgahinga's wardens held a meeting to discuss their forthcoming multiple-use and tourism programs, I jumped at the chance to visit. Liz and I joined Phillip Franks, director of CARE's Development Through Conservation project, for the drive to Uganda's southernmost corner.

Solitary cumulus clouds crossed the blue sky like wind-filled sails as the afternoon stretched towards evening. The road climbed ridge after ridge and snaked through narrow valleys, lengthening our forty mile trip into nearly three hours of undulating, picture-book vistas. Hillside shambas, the crimson bloom of flame trees, a herd of goats: every aspect of the landscape was amplified and sharp-edged, and colors leapt to the eye from great distance, as though the air had never been so clear. People walking the roadside flashed by my window in hair's-breadth glimpses, their features crisp and strangely vibrant—an ancient face, a bright skirt, the teeth of a smile, or fingers tangling shyly in a shock of sable hair. Each image seemed unique and instantly familiar, as if everyone we passed had become suddenly famous.

Like my first views of Bwindi forest or The Great Rift Valley, driving to the Virungas overwhelmed me with landscape. Hillsides rose all around us in an endless patchwork of steep, terraced fields, each their own shades of millet green and maize and fresh-turned earth, framed by the shadow of hedgerows. The volcanos towered higher with every passing mile, their perfect grey-blue cones taking on subtle new dimensions as we closed the distance. A network of finger-thin valleys and dark ridges descended vertically from the peaks, tinged on the lower slopes with the faintest trace of emerald. Rich evening sunlight graced every vista with gold, and it seemed the whole scene was designed to inundate the eyes, to flood all sense and perception with image, image, image.

As the last orange glow of sunset drained from the sky, we drove finally into Kisoro, a languid town of tiny shops and half-finished buildings, nestled in the triple shadow of Muhavura, Mgahinga, and Sabinyo. Like Kabale, Kisoro had suffered from the recent downturn in trade with its unstable neighbors, Zaire and war-torn Rwanda. The main street seemed deserted, and in fact, many businessmen had put their affairs on hold and returned to their family farms, waiting for an upswing in commerce. Phil pointed out the park office, a tidy-looking building on the edge of town, then took us straight to Kisoro's most famous landmark, a small hotel known as The Travellers Rest.

For more than forty years, gorilla-seekers from around the world have launched their Virunga expeditions from this tiny hostel. George Schaller called it a "home away from home" during his seminal 1959 gorilla study, and Dian Fossey took refuge there when her first research site in Zaire was overrun by soldiers. They described a charming, rustic inn surrounded by gardens and flowerbeds, where Walter Baumgartel, the kindly German proprietor, helped arrange camping supplies, guides and porters. Something of a gorilla expert himself, Baumgartel maintained a seasonal camp high on the saddle between Mgahinga and Muhavura, helping pioneer research and tourism in the area. Fossey called him "one of the kindest and most endearing friends I had made in Africa;" or, as Schaller put it: "Walter…greeted us with outstretched arms, his eyes twinkling, a happy smile on his face; and we were content."

Phil pulled in to the hotel's dusty yard and we began unloading the car, but the place looked as desolate as the rest of town, and no one came rushing out to welcome us. Baumgartel had sold the Travellers Rest and left Uganda in 1969, disheartened by the country's rapid decline under Milton Obote. From the dilapidated state of the buildings, it seemed he'd taken much of the hotel's personality with him. We passed his car, a classic 1950s sedan, abandoned on a nearby side street. Wheels gone and body growing through with weeds, it was a true relic, a rusted reminder that for Westerners, the 'romantic age' of travel in Africa has long since faded into memory.

Legendary wilderness and a rich cultural heritage draw more visitors to the Dark Continent every year, but the modern African experience is tainted by a vague sense of desperation. One cannot pass through these landscapes, however magnificent, and not feel the accelerating strain on every system—natural, social, and political. Fueled by overcrowding and volatile leadership, most countries are like Uganda, caught in the struggle to leap from a society of subsistence to a Westernized world of foreign religions, national governments, and market economies. A strange combination of hope and futility prevails, where every change or improvement is overshadowed by the knowledge of its own impermanence. In Kampala, I once watched several men pouring a new concrete sidewalk. They worked steadily, even cheerfully, all afternoon, but poor quality cement and a bad mixture made the pavement split into a spiderweb of fine cracks as soon as it dried. Two days later, large chunks were already sluffing off into the roadway, a simple illustration of a much larger theme: striving for changes that begin to fall apart before the foundation ever has a chance to set.

During the colonial period, *muzungu* travellers saw a much different view of Africa. While the colonial governments were setting terrible precedents for the future—creating arbitrary political boundaries, exacerbating tribalism and forging economies of dependence—they succeeded in erecting, however briefly, a working facade of Europe over an exotic tropical landscape. In Uganda and Kenya, white settlers and their visitors lived in a challenging, but comfortable world, like rural England with better weather. Journalist and author Alan Moorehead described such a visit to Kabale in the 1950s:

*"It possesses a delightful English inn set among lawns and ter-
raced gardens. There is a well-kept golf course just outside the
grounds, and within the immediate neighbourhood of the hotel itself
one can play tennis, badminton, croquet, bowls, table tennis…In
the evening one drinks French wine at dinner, reads the magazines
in the lounge, plays bridge and listens to the radio. Very rightly the
European inhabitants of East Africa take their holidays in this cool
green place, for it bears a striking resemblance to any of the lusher
golfing resorts in southern England, Sunningdale perhaps."*

Expatriates share a certain nostalgia for the old days, and even
Ugandans sometimes longed for the colonial sense of order.

"Everything worked!" Tom had told me once, sitting on his porch in
Kajansi. He was a little drunk, and making grand statements. "Bring the
whites back; give the government to the British! Give the shops back to
the Indians! Then it won't cost two thousand shillings for a lift in a broken taxi!"

I felt a bit wistful myself, checking in to a cold room at The Travellers
Rest, with its broken windows and paint-peeled walls. A dusty sink was
all that remained of the plumbing, and electric bulbs dangled useless
from the ceiling; the generator had ground to a halt years ago from a lack
of machine oil. But soon a friendly old man knocked at the door, bringing candles and a basin of warm water. I thanked him in Rukiga and he
shook his head, "*eh, eh, eh,*" then taught me a proper greeting in
Rufumbira, the local dialect. I washed my face and dried it on a bedsheet,
musing on the hotel's earnest, dilapidated charm. This is the appeal of
travelling in modern Africa, an attraction I heard one tourist aptly call
"entropic hospitality."

Echoes of the colonial period still survive in Uganda, coloring the
attitudes of many expatriates and locals alike, but they seem increasingly
out of place in the modern forum. I met one such anachronism at our

multiple use meeting the following day. As the field director of a German-based gorilla project (BRD),[*] Klaus Jurgen-Sücker held the title of honorary warden in Mgahinga Park. A tall, gaunt man with shoulder-length grey hair and a brooding gaze, Klaus was an imposing figure who dominated the park's Ugandan staff.

"There can be no multiple use," he interjected angrily for the fourth time. "There can be no people in the National Park. They will only drive away the gorillas!"

The debate focused on a CARE proposal to give local beekeepers access to the forest edge, and allow the seasonal collection of young bamboo shoots for replanting outside the park. Public relations around Mgahinga had been in decline since a recent resettlement scheme removed over 1,300 people whose farms had encroached deep inside the park boundary. CARE hoped that a multiple use plan, coupled with a new community water source, would help ameliorate the ill feelings. From a biological standpoint, however, Klaus definitely had a valid point. In spite of the resettlement, Mgahinga still covered fewer than twenty-two square miles, by far the smallest park in the country. Gorillas and other wildlife were sometimes forced to migrate into neighboring parks in Rwanda and Zaire. The dangers of human use, from fires to disease transmission or over-exploitation, increased dramatically in such a limited, fragile habitat.

After hours of heated discussion, the other advisors and wardens over-ruled Klaus's voice of dissent, approving an experimental bamboo harvest for the following month. Everyone agreed that letting public resentment fester unaddressed would set a bad precedent for the young park. In the past, conservation policies focused solely on preservation, often at the expense of local communities. Now, with population growth driving the demand for land and forest resources ever higher, parks can't afford to alienate the populace. In a region as politically turbulent as Central Africa,

---

[*]   Berggorilla und Regenswald Direkthilfe-Mountain Gorilla and Rainforest Direct-Help.

government systems can change rapidly, or disappear altogether for a time, leaving local people as the ultimate stewards of the forest. Without some level of community support, efforts to protect areas like Bwindi and the Virungas run a serious risk of failing over the long term.

An encouraging lesson can be learned from neighboring Rwanda, where the mountain gorillas and Parc des Volcans survived years of civil war and tribal massacres relatively unscathed. The conservation education and public outreach begun in the early 1980s had apparently helped both sides of the conflict recognize the importance of the national park, as well as the economic potential of gorilla tourism. Bwindi and Mgahinga hope to emulate this apparent success, embracing a modern conservation strategy that utilizes every possible tool, from traditional anti-poaching patrols to more progressive ideas like ecotourism, multiple use, and revenue sharing. The newer concepts hold great promise, but no guarantees to placate a traditionalist like Klaus, who left the meeting in frowning silence. I could almost sense his longing for the past.

The following morning, all of Africa seemed to drop away beneath us as we climbed the mountainside for a trip to the gorillas. Lakes Bunyonyi and Mutanda glinted up from their valleys like flattened silver coins, and the ridge-backed hills of Kigezi rose up around them in miniature, faded green in the distance, like the paint-and-plaster landscape of a scale-model train. The sky was exceptionally clear, but white clouds still clung like vapor trails to the crest of Mt. Muhavura, and brushed across Mgahinga's low, flat-topped cone. Mt. Sabinyo, "the old man's teeth," towered above us to the southwest, obscuring the Rwandan and Zairian Virungas behind its distinctive crown of five jagged peaks.

Klaus led us up the slope in long, easy strides. The entire staff lined up to salute him at the park entrance, and a lanky young ranger immediately leapt forward to carry his backpack. I caught Liz rolling her eyes and smiling, but while his militant style bordered on the absurd, no one questioned Klaus's accomplishments, or his dedication to the gorillas. In the five years of his tenure at Mgahinga, the place had progressed from a neglected forest reserve to an organized national park, with more anti-poaching rangers per square mile than any other protected area on the continent. He had led the struggle to evict the park's encroachers, and we now witnessed the results: a full hour's climb through the lumpy, uneven ground of abandoned millet fields and banana *shambas*. After watching the steady destruction of wooded patches around Bwindi, it was deeply satisfying to see a bit of farmland returning to the forest. A small victory perhaps, but Mgahinga may well be the only place in Africa where gorilla habitat is actually expanding.

"My uncle's farm was here," a ranger told me, gesturing towards an overgrown field indistinguishable from the rest of the hillside. I asked him how the family had felt about leaving, and he shrugged noncommittally. The park compensated farmers for their land and lost crops, but resent-ment still ran high in the local villages, and Klaus bore the brunt of it. His conflicts with the community, and increasing disagreements with other conservation groups, cast doubt on the future of BRD in Mgahinga. Local leaders, government officials and even projects like CARE were mounting pressure to cancel Klaus's advisory contract with the park, paving the way for a more cooperative approach to management.

We paused to rest at the edge of the bamboo zone, a tall, vertical tangle of reeds that ringed the mountains above 8,000 feet. Virunga gorillas craved the tender young shoots, and migrated seasonally to take advantage of any new growth. Twenty-five miles to the north, Bwindi's gorillas rarely even entered the bamboo, and had never been recorded feeding on its shoots. The difference illustrates an important point about the two popu-lations. Bwindi gorillas live in a montane rainforest, roaming from an ele-vation of roughly 5,000 feet to the highest reaches of the park, at more than 8,200. In the Virungas, however, all habitat lower than 7,000 feet has been gone for centuries. Villages and farms press right up to the park boundaries at the bottom of the bamboo zone, and the gorillas' range extends upwards, through mist-enshrouded *Hagenia* woodlands, and into alpine areas higher than 12,000 feet.

This distinct difference in elevation and habitat accounts for the disparate diets and ecology of the two populations. Bwindi gorillas must range over a larger area, moving long distances between forest clearings to ensure a steady supply of lush vegetation. The dense shade of mature rainforest screens out many of the moist plants that gorillas rely on, but they do take advantage of its greater variety of fruits, leaves and tree bark. Wild figs (*Ficus* spp.) and the pineapple-like Omwifa (*Myrianthus* spp.) are common meals for Bwindi gorillas, but unavailable to their volcano counterparts. In the Virungas, gorilla habitat includes bamboo, alpine meadows, and a more open-canopied forest, with an abundance of wild celery and other herbaceous foods. The apes need not travel so far to forage, and their home ranges are correspondingly small. Physical variations separate the two groups as well. After three years in Rwanda and more than a year in Bwindi, Liz could recognize photographs from either population, pointing out the broad faces and noticeably longer hair of the Virunga apes.

The trackers motioned for silence as we pressed ahead into the bamboo, stooping to crawl through narrow tunnels and gaps in the dense thicket. Overhead, the long grassy stalks and narrow leaves rattled quietly in an alpine breeze, casting down a gridwork of shadows and stripes of sun. The gorillas had been feeding nearby yesterday, and with the bamboo shoots in full season, they probably hadn't moved far. Twenty minutes later we found the site of their morning meal, an open place scattered with fresh peelings, and bits of turned earth. I picked up a leftover shoot and tasted it: watery and slightly bitter, like the rind of a cucumber. The trail led on, but the gorillas were close now and we moved more slowly, single-file through the dimness.

Suddenly, we heard a long sigh, answered by the quiet grunt of a gorilla at rest. These vocalizations come from a kind of deep, controlled belch, and usually indicate a mood of relaxation. Humans can mimic the sound with breathy coughs, and we often did so to announce our presence, or send calm messages to a nervous group. I heard Liz clear her throat, and was about to do the same when one of the trackers let out a low, crooning whoop. Liz and I exchanged a startled glance, as the other trackers joined in, "Woooop. WooooOOOP!"

"They sound like hyenas," she whispered, and we stifled a laugh.

The gorillas, however, didn't react at all. Hyena whoops had been a part of their routine for months, a familiar sound that helped them identify their daily human visitors. Across the border, Zairian rangers clapped their hands when approaching a group, and while in Bwindi we had always concentrated on imitating the gorillas' natural sounds, I realized now that consistency was probably far more important than giving an accurate mock-belch. Anything repeated regularly would have the same effect.

We moved in closer and crowded together, peering through small openings in the foliage. The gorillas lounged nearby in a patch of deep shade, grouped loosely around the silverback. There were two females visible, and a pair of juveniles that tumbled lazily over one another, biting and wrestling through the undergrowth. Their play hoots were barely audible, like faint panting or the chuckle of infants. I glanced at Klaus, and saw him smile for the first time. Later, he would invite us to his small house in Kisoro, and recount the day with unexpected laughter over bottles of German beer.

We stayed with the gorillas for close to an hour, and they seemed perfectly oblivious to our presence. Habituated several years earlier in Zaire, the group now spent most of its time on the Ugandan side of the border, and would begin receiving tourists in a few months. I watched them closely through binoculars, and immediately noticed the contrasts that Liz had mentioned. The silverback had a much wider face than Mugurusi, Ruhondeza, or any of the Bwindi apes. His noseprint looked more pronounced, with deep wrinkles, and ridges of black skin that shone as if constantly wet. All the apes appeared shaggier than their Bwindi cousins, particularly the youngsters. With the long hair swept back from their foreheads in wispy strands, they looked like mad scientists, or a pair of middle-aged folk singers.

These subtle variations in appearance confused primatologists for decades. Experts from Schaller to Fossey all mistook Bwindi apes for members of the eastern lowland subspecies (*G. gorilla graueri*), until a DNA comparison showed the Virunga and Bwindi populations to be nearly identical genetically. A continuous band of forest connected the two areas until only 500 years ago, when agriculturalists first settled and cleared the valleys around Kisoro. Before then, mountain gorillas were likely part of a single population that covered a much wider range of altitudes. Zoologists still argue whether or not the two populations should be classified as distinct subspecies, a typical confrontation between the 'splitters,' who divide the natural world into narrow categories, and the 'lumpers,' who use a broader taxonomy. In terms of conservation, the point is moot: they are either mountain gorillas, the rarest great apes in the world, or they are Bwindi mountain gorillas and Virunga mountain gorillas, the rarest great apes in the world. Today's intensive efforts to preserve them would remain the same.

Less than 100 miles away, eastern lowland gorillas inhabit the foothills and forests of Zaire's Kivu region. Apparently, the drier climate of the rift valley floor has acted as a barrier to gorilla migrations, isolating lowland populations from their mountain gorilla cousins long enough for the two to 'drift' genetically into distinct subspecies. Between two and five thousand eastern lowland gorillas survive, many of them in rugged highland areas similar to Bwindi.

The third subspecies of gorilla, the western lowland (*G. gorilla gorilla*), make their home in the sweltering jungles of the western Congo basin. During glacial periods in the late Pleistocene, Africa's climate dried out periodically when atmospheric moisture became trapped in the expanding polar ice caps. The great rainforests of Central Africa shrank into small pockets, separating western gorillas from their eastern counterparts by nearly 1,000 miles of dry savanna. Although the forest has long since expanded again, gorillas have been slow to recolonize the area, and vast distances still separate the two lowland subspecies. Short-haired and smaller in size, the western population has adapted well to their hot, humid environment. More than 20,000 still roam the forests of Gabon, Congo, Zaire, and other West African countries, but they face increasing threats from hunting and large-scale logging throughout the region.

As the first gorillas 'discovered' by the outside world, western lowlands still make up the vast majority of zoo and museum specimens. A zoo in Antwerp, Belgium, supports the only captive population of eastern low-land gorillas. No mountain gorillas have ever survived long in cages.

Moments before we started back to camp, the silverback suddenly roused himself, tilting his body upright in the gloom like an oversized mastiff. He shambled away from us down the slope, and the rest of his family rose to follow, the youngest juvenile clinging to its mother's back, little fists knotted firmly in the long, dark hair. Days later, the group ventured out into the open, crossing several hundred yards of abandoned fields to reach an isolated patch of bamboo, remnants of a village farm slowly returning to the forest. Within a year, the gorillas would reinhabit over eighty percent of the reclaimed land, feeding and even sleeping in the open meadows, as if enjoying the view of Southwest Uganda spread out like a map before them.

Chapter XII

Friends

▼

*"I think I could turn
and live with
animals...They do
not sweat and whine
about their condition.
...Not one
is respectable or
unhappy over the
whole earth."*

—Walt Whitman, *1855*

On my birthday, John treated me to dinner at Hope and Phenny's place, the H&P Canteen. They lived and cooked in a narrow, metal-roofed building directly across from the campground, serving up made-to-order meals of cabbage, beans, fried potatoes and groundnut stew to the growing stream of tourists now visiting the park. John and I were regular customers too. The food was good, but we came just as much to socialize with the couple we'd grown to like so well: Phenny, the prankster with a serious heart, and Hope, a tireless source of kindness and capability in any situation.

In many ways, they represented the best possibilities for a place like Buhoma, two people from traditional families who instinctively grasped the potential of every new development in their community. They held good jobs, built a thriving restaurant business, and moved comfortably between village life and the emerging tourist culture of the park. Hope and Phenny seemed completely at ease with *muzungus*, yet still held respected positions in the village. She helped found the women's group and kept books for the community campground, while Phenny's leadership and work ethic made him a natural successor to his father, a leading village elder, or even his uncle, the local chief.

John had ordered chips ahead of time to go with the bottle of ketchup he'd found in Kampala. We sat inside and Hope came to join us, frowning thoughtfully as she sampled the exotic condiment. "It's a bit sweet for potatoes, John," she concluded. "Maybe as dessert?"

A young cousin came in from the kitchen and Hope directed her on preparing the rest of the meal. She may have been taking a break, but there was no question about who was in charge. Hope was eight months pregnant and bothered by a lingering case of bronchitis, but she still worked full time for the park, ran the Canteen at night, and seemed involved in just about everything that happened in the village.

"I think that Buhoma will just stop when you have your maternity leave," John joked with her and she laughed until she coughed. "I'm serious," he insisted. "You're running this whole village!"

She protested, and excused herself to the kitchen in embarrassment, but John was right. It was hard to imagine a functioning Buhoma without Hope.

After dinner, Phenny, John and I walked deep into the forest without flashlights, stepping carefully along familiar trails made strange by the darkness. We stopped at the first Munyaga bridge, three friends sitting over a black rush of water, talking about where our lives had crossed. I thought of a simple goal I'd had in joining the Peace Corps, to make friendships unbounded by the gaps of cultural difference. With Tom

Ntale I'd gotten off to a strong start in Kajansi, and realized now that my real birthday present lay in finding the same thing happening in Buhoma.

A sudden downpour drenched us on the way home, and I woke up coughing the next morning. Two sneezing fits on the way to the office confirmed it: I'd caught my first rainy season cold. As if sensing my condition, and knowing it would prevent me from tracking, Katendegyere group immediately did something dramatic.

"It is Makale," Phenny told me worriedly. "He fought terribly with the big silverback, and is now very sick. The trackers think he is dying."

No one witnessed the fight itself, but traces of nervous, diarrhetic dung along the trail told them that something was wrong long before they reached the group. Soon they came to a wide, trampled area spotted with blood and bits of torn hair. They found the gorillas soon after, but the group was moving quickly away from the scene of the battle, and they only caught glimpses of the two combatants: Mugurusi bleeding from scratches on his neck and arms, and another male trailing behind with deep head wounds and a mangled wrist.

"Are they sure it was Makale?"

"It should be," Phenny affirmed, but identifying individual apes was still a new concept to the trackers, and the name Makale, 'fierce one' could easily apply to any gorilla involved in a fight.

Every afternoon they came to my house with more bad news. His dung was bloody and he wasn't eating. He couldn't use his injured hand. He made his night nest further and further away from the rest of the group. I told them to keep watching carefully and try to confirm the gorilla's identity, while I stayed at home, cursing the virus and waiting for my symptoms to fade.

Even in perfect health, however, I knew there was little I could have done for the injured ape. Liz had treated numerous gorillas in Rwanda for snare wounds and other human-induced problems, but while we kept tranquilizers and darting equipment on hand, we avoided interfering in the natural hazards of gorilla life. To do so could upset the group's social hierarchy, not to mention the normal process of death and renewal. If Mugurusi had successfully defended his dominant role, sending a rejuvenated challenger back into the ring invited an unnatural conflict and behavioral disturbance.

Finally, I woke with a clear head and no trace of coughing. The gorillas were nearby, lingering in a narrow canyon high above the Muzabajiro valley. Prunari told me they hadn't moved far since the day of the fight, a typical behavior for groups with wounded or sick members. They adjusted their daily routine, travelling short distances and allowing their wounded companions extra time to feed.

"Makale is eating again," Charles said, stopping on the trail to make the sign of a full belly with his hands. But the ape was still very shy, Phenny added, staying in the thick brush a good distance away from the rest of the group.

When we reached them, the gorillas had spread out feeding in the dense undergrowth, visible only as a group of indistinct shadows and leaf-tremors. We climbed to a small ledge at the head of the valley, hoping they would pass through an open place below us. The trackers prided themselves on choosing good vantage points, and predicting the gorillas' movements. When the first ape appeared in the clearing, Prunari looked back at me with a self-satisfied smile.

Suddenly, the gorilla lunged up the slope towards us with an angry scream, pulling at clumps of vegetation and baring his teeth. Gorillas usually prefer to charge from above, but this was Katome, "the small fist," Makale's little brother and the youngest of the black-back males. He often bluff-charged to cover for Mugurusi, and sure enough, the "old man" darted quickly across the clearing behind him as we backed away. I caught a glimpse of the red gashes on Mugurusi's forearms before he disappeared up the hill. When gorillas fight, their long canine incisors can inflict terrible damage, and while Mugurusi may have emerged the victor, he obviously hadn't escaped unscathed.

Katome fixed us with a beady-eyed stare as he withdrew, turning to follow Mugurusi up the slope. We rarely saw him venture far from the aging silverback, and before they settled on a name, the trackers had referred to him simply as "the assistant." All the male gorillas in Katendegyere group were probably related, either the sons or younger brothers of Mugurusi. Lead silverbacks often tolerate several other males in their group, but they will aggressively defend exclusive mating privileges. Conflicts arise when the younger males attempt to breed, so most choose to emigrate when they reach sexual maturity. Travelling on their own, they gain strength and experience, hoping to attract females and start a new family. Some apes, however, remain in their natal groups for life, aging gracefully as secondary, non-breeding silverbacks. Judging from his close relationship with Mugurusi, Katome seemed bent on following the latter path.

Several minutes later, Nyabutono crossed the clearing, followed by her three-year old child Kasigazi, "the little guy." The rest of the group stayed out of sight behind forest's curtain of leaves, but we heard them slowly making their way uphill, until only one ape remained below us, feeding quietly in a dense thicket.

Phenny motioned towards the rustling bush with his lips, a distinctly Ugandan gesture developed to avoid direct pointing, which is considered rude. "This should be the one," he whispered.

We settled in to wait, hoping the gorilla would choose a visible route when he decided to follow his family up the slope. Twenty minutes later, the other gorillas were completely out of earshot when he finally began to move. We watched the brush part around him as he emerged into the clearing, walking gingerly, with his left hand twisted backwards at a sickening angle. A deep cut ran from behind his ear to the top of his head, laying the flesh open to the bone. At this distance I couldn't make out his noseprint, but when he turned sideways, a band of silver hair showed clearly across his lower back.

"*Rugabo*," I whispered to Phenny, and handed him my binoculars.

"Yes," he agreed, "a silverback."

Not Makale, I thought, and was surprised by the strength of my relief. Since hearing of the injury I'd grown increasingly tense with worry, as over a sick friend one can do nothing to help. I was surprised too that the trackers hadn't recognized this gorilla as the third silverback, the only unnamed member of the group. But he had always been reclusive, and we didn't know him nearly as well as the others. I watched now as he made his way slowly across the far edge of the clearing, stopping twice to rest and feed. By pulling down branches with his good arm and bracing them with the broken hand, he could easily strip off the edible bark and leaves. If he managed to avoid infection, a full recovery seemed likely. The skeletons of wild gorillas often show signs of healed fractures and other serious injuries. Fossey found old head wounds in nearly three quarters of the silverbacks she studied; two of the skulls actually contained the cusps of their adversaries' canines, broken off and fused into the mended bone.

After the gorilla had moved out of sight, I turned to the trackers. "This one we will always know," I said, holding my left hand bent back at the wrist.

"*Kacupira*," Prunari supplied immediately: "the one who limps."

The name completed our lexicon of Katendegyere group, eight individuals as distinctive as any of my other friends and co-workers. Realizing this, I thought back to training, and my first real gorilla behavior discussion, with a noted expert named David Watts. *This is a little odd*, I remembered thinking, as he continually referred to the apes with human pronouns: "someone grunted" or "everyone came over, and I know they remembered me." Now I was doing the same thing, not to mention my degenerating feeding habits, and a growing tendency to clear my throat with a cough-bark. When I started making night nests and raiding Behuari's banana *shamba*, I would know it was time to go home.

((((

"You're from Washington?" she cried and turned to find her husband. "Honey, come meet…Todd. He's from D.C.! Get the camera!"

Her husband waved back without answering. He was cheerfully sharing cigarettes with the trackers and chatting about the lions he'd seen in Kenya. Charles and Prunari looked hungover, but smiled their thanks, stowing the cigarettes carefully in a dry pocket of their tattered backpack.

"What are your names? What can I call you guys?" I heard the man ask.

Charles stared back in vague comprehension, then gave him the baffling answer, "God made me."

"Great!" The man nodded enthusiastically, and I tried not to laugh. Good natured non-communication always set a jovial tone for the day, and I just hoped we could avoid more serious complications. A week before, the tourists had all gone hungry after their porters mistakenly devoured the lunches they'd been hired to carry.

With three new safari camps in Buhoma, more high-end American tourists were visiting Bwindi Park. People reacted with surprise on finding a Peace Corps volunteer in the jungle. Younger travellers often envied my position, but anyone with children echoed the response of today's group: "Your parents must be worried sick."

Half an hour later, all conversation came to a halt on the steep slopes of Rushura. The couple and their friends were in their mid-sixties, on an extended East African trip sponsored by a Midwestern zoo. After two weeks of touring the savanna in mini-vans, hiking up a mountainside came as something of a shock. The guide and trackers stopped frequently and kept the pace slow, but it's still a difficult climb, and the eldest of the men started falling behind. I watched him stumble and lean into the hillside with every step, sweat pouring from his forehead.

"He had a triple by-pass last March," his wife confided to me behind her hand.

"Ah," I answered and shouted up to the trackers: *"Turuhuka Baa Ssebo!!"*—'We rest!'

In Bwindi, I learned that determination can overcome almost any exhaustion, and rarely saw tourists turn back without reaching the gorillas. But our first aid kit stopped somewhere short of defibrillation paddles, so it was a great relief when the trackers stopped climbing and veered south, descending into a shady side-valley.

We made good time using an old poachers' trail through the forest. These faint paths criss-crossed the hillside like a network of secret highways, and the trackers knew them well. Before working for the park, many of our staff had found other employment in the forest, hunting duiker and bushpig, or cutting mahogany. By hiring poachers and loggers, we not only gained the best woodsmen in the area, we cut down on illegal activities by providing them with an economic alternative.

Using old trails also minimized our impact on the rainforest. The trackers avoided cutting anything unnecessarily, but following a gorilla group's wandering route would have been impossible without a sharp *panga*. Luckily, gorillas preferred thickets and clearings where the vegetation grew back quickly, erasing the signs of their passage and ours within a few short weeks.

The valley narrowed into a rocky cleft, and we clambered over wet stones before the trail bore upwards again through the green-dark shade. Thunder drummed across the hidden sky, and rain drifted down around us in a cool, enveloping mist, as if the very trees were steaming.

Suddenly, the trackers stopped short, their *pangas* in mid-swing. I heard a faint rustling somewhere ahead of us, and Prunari stared intently into the shadows, listening. "*Enkima,*" he whispered finally, and a troop of L'Hoest's monkeys appeared, quietly crossing the path in front of us. Silvery-grey, with chestnut backs and narrow, white-bearded faces, they moved like quick ghosts, disappearing back into the undergrowth without a sound. Also known as the mountain guenon, L'Hoest's were the rarest of Bwindi's five monkey species, their range limited to a handful of high-altitude forests in the Central African region. The tourists whispered excitedly amongst themselves, scanning the foliage for another glimpse of the monkeys, and we continued on in hushed silence, with a newfound sense of expectation.

Soon after, the trackers found fresh gorilla-sign, and we turned to climb a familiar slope. Katendegyere group ranged over twenty-two square kilometers of jungle, but their pattern of movement took them to a series of predictable foraging grounds. Before we even heard the cough-bark and pig-grunts of a feeding dispute, we knew the apes had gathered around their favorite, termite-ridden log. We stopped several hundred yards away to let the tourists prepare their cameras, then left the extra gear behind with the porters, advancing slowly in the smallest possible group.

Prunari led us in a wide circle, bringing the gorillas into view directly below, gathered loosely around the bleached crown of a huge old windfall. The snag had blown down across the slope, and lay in three long sections, slowly decomposing into the leaf-and-twig litter of the forest floor. I counted six gorillas feeding on the termites. Karema sat to one side, peering intently at a strip of orange bark she'd torn from the tree, and nibbling the helpless insects as they crawled out into the light. Kacupira and Katome were busy ripping their own meals from the stump of a wide limb, while Mugurusi lay sideways with his head completely out of sight beneath the fallen trunk, audibly gnawing at the rotten wood to reach the white ants below.

Unlike chimpanzees, gorillas have never been observed using tools in the wild. Where the smaller apes craft particular twigs to 'fish' for termites, gorillas use brute strength to break apart the insects' chambered mounds or rotted wooden homes. As the more arboreal species, chimps have developed greater dexterity, enabling them to balance and feed simultaneously, high in the forest canopy. Gorillas spend most of their time on the ground and haven't learned to manipulate tools, but researchers regard them as equally intelligent, with a greater capacity for patience and imagination in their problem-solving techniques. Both species have been able to master sign language in captivity, communicating to each other and their human observers with vocabularies of more than four hundred words and phrases.

Chimpanzees have even learned to teach these signs to their offspring. In human terms, the intellect of great apes is often equated to that of a three or four year old child.

Mugurusi's three year old, however, still had a lot to learn about the fine art of termite hunting. I watched Kasigazi climb up his father's broad back to perch near his shoulders, gazing curiously under the log where the old silverback was feeding. Mugurusi grunted, and shifted himself into a better position, knocking the young ape to the ground. Kasigazi scampered over to his mother, and stood behind her, hooting plaintively with a piteous expression that communicated clearly in any primate language: "*give me some ants; give me some ants.*" But Nyabutono ignored him completely, picking termites from a deep furrow in the log and chewing them individually with great relish.

Watching this peaceful scene, I had nearly forgotten the missing pair of gorillas until a sudden commotion erupted somewhere up the slope. Screams and pig-grunts resounded through foliage as Makale and Mutesi squared off over some imagined quarrel. The two had been inseparable for months, moving on the periphery of the group. I expected one or both of them to be the first males to emigrate from Mugurusi's bachelor band, but so far they had been satisfied showing their independence by mock-charging, and carrying on against one another. I trained my binoculars up the hill, watching their conflict unfold like a shadow play of indistinct images in the undergrowth.

When aggression between gorillas reaches the boiling point, the steam often looks for the closest, easiest outlet. So I wasn't surprised when, after several minutes of fierce vocalizations, Makale and Mutesi charged rapidly downhill, directly toward the smallest primates in the immediate area. We gathered the tourists into a tight group and crouched down, as the gorillas crashed through the intervening vegetation.

Suddenly, Kacupira appeared, sprinting up the slope to place himself between us and the advancing apes. With his maimed hand, he couldn't hope to challenge two healthy males, and immediately assumed a submissive posture, holding his hind quarters high in the air. Makale and Mutesi screamed indignantly, but Kacupira had diverted their attention long enough for us to back slowly out of the vicinity. After several tense minutes, he rose and rejoined the rest of the group. Makale and Mutesi seemed mollified by this humble display, and calmly followed him down the hill without so much as a glance in our direction.

I looked at the trackers in mute wonder and they shrugged back with smiles of amazement. Later, when we stopped for lunch, one of the tourists replayed the incident on his video camera, and I watched Kacupira defend us again and again. I had never seen a gorilla behave that way, not only acknowledging our presence, but reacting to it as if we were members of the group. Usually, maintaining our distance minimized any behavioral impact on the gorillas, but today we had clearly influenced their social dynamics, and for Kacupira, perhaps even his thoughts and perceptions.

Researchers and zookeepers have known for decades that gorillas can recognize individual people. Captive apes clearly remember their favorite observers and trainers, even after years of separation. In wild populations, habituators learned that the gorillas should be exposed to a mixture of light and dark-skinned people. Otherwise, the process suffers a serious setback when the gorillas react with fright to the sudden appearance of *muzungu* scientists or tourists among their daily visitors. In Katendegyere group, the apes always noticed when Prunari, Charles or James went on leave. We would often recruit one of the anti-poaching rangers as a fill-in tracker, but it soon became difficult to find volunteers. Twice in a row, Makale circled around the rest of us to charge directly at 'the new guy,' as if testing his mettle for habituation work. Once, he even reached out and laid his hand on the cowering ranger's shoulder before turning to stalk smugly back into the undergrowth.

"The gorillas have finally accepted me as one of their own!" I joked in a letter home, describing the encounter with Kacupira. But as our knowledge of individual apes grew more intimate, I wondered how much impact they

were having on my own behavior, and that of the trackers. We knew to
tread lightly around Makale, and we knew that Mugurusi was shy. Now we
felt closer to Kacupira, referring to him with affectionate Rukiga terms like
munwani waawe, 'our friend.' As the gorillas became more comfortable
with us, we went through our own set of subtle changes, leading inevitably
to the puzzling question: who was habituating whom?

CHAPTER XIII

LAST RITES

▼

When he was here
We joked and laughed together
And no fleeting shadow of a ghost
Ever crossed our paths.

—Laban Erapu, Ugandan poet
from "An Elegy"

In the late spring of 1994, Uganda's commercial fishing industry collapsed. Throughout the country, restaurant and hotel owners erased Nile perch and tilapia from their menus, and the busy lakefront markets grew quiet.

"No fish," I observed, rattling along in a crowded bus towards Kampala. Along the roadside, dozens of small wooden stalls stood empty where fishmongers normally hawked their wares.

"Who would buy it?" asked my seatmate. "You could be eating a neighbor, or a relative."

He referred to a new, unthinkable component in the Lake Victoria food chain. For over a month, bloated human bodies had been floating down the Kagera River, drifting into the lake and washing ashore on Ugandan beaches at the rate of fifty or a hundred every hour. The army, aided by relief agencies, set up wide nets along the delta, and patrolled the river in boats, but thousands of corpses still slipped by, victims of the civil war and genocide raging across neighboring Rwanda.

When extremists in the country's Hutu-led government shot down the plane of their own president, Juvenal Habyarimana, they derailed his fragile peace negotiations with the Rwandan Patriotic Front (RPF), a rebel group dominated by the minority Tutsi tribe. The assassination rekindled a three-year old civil war, and as RPF forces advanced in the countryside, the government retaliated savagely, orchestrating the systematic slaughter of more than 500,000 innocent Tutsi civilians. Carried out by neighborhood militias and retreating Hutu soldiers, these massacres marked the worst killing in a struggle dating back more than forty years, rooted in ancient tribal animosities, and the legacy of colonial rule.

A tall, pastoral people of Nilotic origins, the Tutsis migrated into Rwanda hundreds of years ago, bringing their cows and a governing system of local chiefs ruled by a monarch of god-like stature. They soon dominated the area's resident Hutus, a race of shorter, Bantu-descendent agriculturalists. The social hierarchy that developed has been compared to a feudal system, where the Hutus provided farm goods in exchange for the military protection of their Tutsi lords. Ethnic lines became blurred, however, by extensive intermarriage and mobility between the classes. The tribal distinction almost became one of occupation as much as race. A rich Hutu who bought many cows might then be considered a Tutsi. Similarly, any Tutsis who lost their herds and took to farming would also lose social status, and become known colloquially as Hutus.

After Belgium took control of Rwanda in 1916, tension between the groups changed quickly from a class struggle to one based strictly on race. Choosing to rule through the Tutsi aristocracy, the colonial government issued tribal identity cards, dividing the population along clear and uncompromising ethnic lines. Similar to South Africa's infamous apartheid system, the Belgian method reserved specific privileges for each race. Education, political power, and economic opportunity became the exclusive domain of card-carrying Tutsis, who made up less than fifteen percent of the populace. In the decades that followed, tension turned to violence as Hutus struggled for equal treatment, and the Tutsis fought to maintain control.

The first major uprising in 1959 overthrew the Tutsi monarch and sent thousands of his tribesmen into exile. Since independence, Hutus have controlled the government, violently suppressing the occasional Tutsi insurrection. The latest rebel movement, the RPF, involved numerous expatriates who had never even lived in their home country. Many had grown up in Uganda, where they received military training and tacit support from the Musevini government. Political cartoons in Kampala played upon stories that Musevini's mother was of Tutsi descent, and criticized him for fighting a 'secret war.' People said he was repaying a debt to the Tutsis, who were rumored to have helped his own rise to power in 1986. But most Ugandans regarded the atrocities in Rwanda with a kind of knowing sympathy, mixed with relief that in this decade at least, civil war and racial slaughter were no longer on their side of the border.

We reached Kampala in the dusty heat of evening, with the sun hung like a lurid Christmas bulb over the western hills. Once described as a picturesque town of neat gardens and tree-lined avenues, Uganda's capital now bore the scars of its turbulent post-colonial history. Architecturally, the downtown area was a disorganized jumble of colonial-style buildings mixed together with towers and apartments from the mid-twentieth century 'cement shoe box' school of design. Bullet holes and years of neglect still marred many structures, and the roadside flowerbeds were littered with barbed wire and heaps of rubble, or planted with clumps of yams. But ten years of political stability, combined with the continent's fastest-growing local economy, had stretched a new feeling of vitality across the city's worn facade. Half-finished construction projects, abandoned for decades, now hummed with activity, and fresh blacktop smoothed the potholes of every major street. The population had swelled to nearly a million, as people from the countryside arrived to fill the suddenly booming job market.

I walked uphill from the bus park through a rushing crowd of commuters, past street vendors who spread their merchandise out before them on the dusty pavement: soap, pens, batteries, bread, radios, foam mattresses, underwear, eggs, and bags of milk. As the daylight faded they would light tiny oil lamps, transforming the street markets into a sea of constellations, glittering like an urban version of starlight. At the corner of Kampala Road, the sidewalk changed suddenly to shiny white tile and I looked up—what had been a derelict building months before was now a newly-remodeled international bank, gleaming with marble and tinted windows. Kampala was changing so quickly, it seemed like a different city every time I came in from the forest. Only a year before, walking through downtown after dark was an eerie experience of deserted, ghost-quiet streets. But with the addition of lighted sidewalks, three major hotels, and dozens of new restaurants, the town had quickly developed a thriving night life. Perhaps the most telling evidence of change dangled above the busy intersection of the Entebbe and Kampala Roads: the city's first stop light. No one paid any attention to it, of course, but it remained, gamely encouraging the traffic with flashes of green, yellow and red like a constant reminder of things to come.

Heading towards the post office, I heard a familiar name shouted over the traffic noise.

"*Jjuko*!!"

Turning towards the voice, I spotted D.K. making his way through the crowd of pedestrians. This was another feature of my trips to Kampala: I always ran into a friend of Tom's who knew me from Annette's place.

He shook my hand excitedly, and asked when I was coming back to Kajansi.

"Tonight," I told him. "I only have these few things to do in town."

Thirty minutes later I boarded a *matatu* and sped through the suburbs towards my old neighborhood. D.K. had gone on ahead, so I knew that Tom would be expecting me at the club. I reached Kajansi under the fading port-stain glow of a dry season sunset, and ducked into the warren of back alleys, tracing my way to Annette's. The courtyard erupted with a chorus of "Jjuko!" and everyone stood up to shake hands, but surprisingly, Tom wasn't among them.

"We will go to his," said D.K. "It is no problem."

Annette poured two small jerry cans of *munanasi*, and we left in a group, singing the Luganda favorite, "I'm so very happy," in loud, boisterous voices. Tom heard us coming and threw open the door of the house.

"My son!!" he cried, hugging me with one arm as he ushered everyone into the sitting room. Sam appeared with a tray full of glasses, and Tom began pouring the wine. I noticed immediately that something had changed about him; he'd lost weight, perhaps twenty pounds. In the States, I would have complimented him on getting in shape, but for Ugandans, weight loss carried a different set of connotations. To be fat signified financial prosperity and good health. People only lost weight if they were sick or going hungry, and I knew Tom was getting enough to eat.

"To *muzungus* in Kajansi!" he made the toast grandly, and we all raised our glasses. "This year I will host two volunteers, or even three!"

We drank wine until late into the evening, and I passed around photographs of Bwindi—my house, the forest, the gorillas. Everyone laughed that a *muzungu* would travel so far to live in such a hut, and chase wild animals in the jungle. When the *munanasi* was gone, people rose to leave, but Vincent had "gotten a little excited," and slowly slipped out of his chair onto the floor. Tom and I bid the others goodnight, rolled the drunkard onto a spare mattress, and joined Susan in the next room for dinner. As Sam knelt to rinse my hands, I saw Tom carefully shaking pills from a small paper envelope.

"I have had malaria," he confessed. "And now typhoid. This is my second treatment, but I'm still a bit weak."

He assured me that his health was improving, but at breakfast the next morning he looked haggard, as if he'd hardly slept.

"Pah," he said, pushing away his plate of yams in disgust. "No appetite."

We walked through town together, greeting acquaintances along the path behind the market. At the Entebbe road, Tom stopped beside his favorite *duka* and the old proprietress brought out a stool.

"The Ministry car pool will pick me from here," he said, "but there are many taxis for Kampala." He stepped to the edge of traffic and waved down a *matatu* for me, swinging his briefcase like a flag.

I climbed aboard, thanking him for his hospitality and promising to visit again soon. As the van pulled away, I watched Tom wave goodbye, then turn to greet a co-worker, and take a seat in the shade to wait.

Two months later, I saw Tom Ntale for the last time, dying in a crowded ward at Nsambya Hospital. I heard the news from Vincent, who spotted me on a busy Kampala side-street.

"Your friend Tom is very sick," he said, shaking his head with concern. "Very sick. You should visit him soon."

I shared a ride to the hospital with Steve Ulewitz, a volunteer I'd trained with in Kajansi, and his homestay father, a close neighbor of Tom's named Charles. As the taxi wove through traffic, they talked excitedly about African teams in the World Cup, and the small television they'd bought to watch the finals. But we all grew quiet when the van dropped us at the crowded stop for Nsambya.

The hospital covered several acres, and for half an hour we wandered along dusty paths from wing to wing, scanning the faces of the sick. My anxiety for Tom turned to angry frustration as the nurses shrugged off our questions. They kept no central registry, and the place seemed designed to confuse people, to make one lose hope of even finding your way, let alone being healed.

Finally, Charles spotted someone he knew, and hailed the man from across a weedy courtyard. We shook hands and made introductions before asking him which way to go.

"Yes, yes. Tom is in the St. Gonzaga building," he told us, pointing back from the way he'd come. "It is a very nice one," he added, looking earnestly at me and Steve. "Bed number 24."

We thanked him and walked on to a low, tile-roofed structure near the back of the compound. Simple cots lined the walls in rows of three, and the patients sat or lay prone quietly, with their families gathered around them. In Uganda, hospitals provide a bed and medicine, but leave food, bathing, and other nursing duties to the patient's friends and relations.

I tried to brace myself for the worst as we counted our way to the twenty-fourth bed, but nothing could have prepared me for Tom's grim condition when we pulled back the curtain around him. I glanced at the emaciated body and thought we'd been misdirected, until I saw Susan on the floor nearby, knitting something from a skein of pink and green yarn.

Tom lay on the cot as if collapsed by the weight of his blankets, eyes closed, a slowly-breathing skeleton. Beyond gaunt, his face was almost unrecognizable, a sharp-edged skull stretched tight with pallid, unhealthy skin. The shaggy beard and most of his hair were gone, and it took careful scrutiny to make out an outline of the proud, laughing man I knew him to be.

I watched his eyelids flutter as we greeted Susan, and she sat up to whisper something in his ear.

"Jjuko," he breathed in a voice like dry leaves. I leaned in close, and his eyes focused slowly on my face. "Why have you come to see me like this?"

I had no words, and only reached down to hold his bony hand. "I wanted to see you, Tom," I managed at last, forcing a smile.

He nodded, but looked suddenly exhausted by the exchange. Susan helped him roll over, facing away from us to rest. I asked her about medicine, and pressed a handful of bills into her palm as a new group of visitors edged around the drawn curtain.

Aunt Florence, Elvis, and several neighbors smiled and whispered greetings to me, shaking hands over the foot of the bed. I had the feeling we were like archaeologists, gathered to view a famous hieroglyph or cave painting before it disappeared, fading slowly from exposure to the wind and the damp breath of our own voices.

It was crowded now, and Charles raised his eyebrows, ready to leave. Tom appeared to be sleeping, but as I shook Susan's hand goodbye, he called for me again.

"We have lost a friend from the club," he whispered as I bent towards him. His lips worked dryly over his teeth, and he looked bewildered for a moment before continuing. "You knew her. Annette."

"I'm very sorry to hear," I answered.

"Yes," he said, turning away again and shutting his eyes, too weak to say anything more.

Outside, the world seemed unnaturally bright; the scent of wood smoke and particles of dust hung suspended air as if trapped in amber and speared alive by sunlight. Particular sounds leapt out of the city noise in quick succession, normal but somehow chilling: a distant car horn, laughter, radio static, or the maniacal chatter of weaver birds. We hurried towards the taxi stand without speaking.

"It is very serious," I commented finally.

"Yes. He cannot recover from this," Charles pronounced each word with grim certainty. His older brother had died the same way, as well as "too many neighbors to count."

Honking *matatus* pulled up to the curb in a steady stream, but the thought of crowding in with so many unknown bodies was suddenly repellent and overwhelming. I said goodbye to Steve and Charles and set out on foot, almost running, as if physical exhaustion might overpower my sadness.

Head down, I walked quickly through the market towards Gaba Road and Liz's distant city house, ignoring the stares of passersby and their constant calls and hoots: "*Muzungu, where are you going? Muzungu, ssst! Muzungu, muzungu 'we!*"

One cannot live for any length of time in Uganda without losing friends to AIDS, but seeing Tom's devastation, so ravaging and complete, left me numb with its implications. Susan was now at risk, and perhaps even the youngest children, Rita and John. With so many people infected, I wondered about the future of a town like Kajansi. "*Everyone you see here is carrying the virus,*" Tom had told me. "*Everyone.*" I knew he exaggerated. I knew that people would survive, just as they survived Amin, Obote, and twenty years of anarchy. But not without a long and intimate re-visitation from sorrow, enough concentrated grief to taint the very ground.

My headlong pace brought me finally to Kansanga, a large trading center where throngs of people milled through the marketplace and hawkers shouted for business outside the shops. The evening rush hour sped past in clouds of dust and diesel smoke, and I stopped to drink a cold soda. Two hand-written signs hung from eaves of my favorite local restaurant.

"FISH TODAY. FISH—NO PROBLEM."

Ugandan menus once again featured tilapia and Nile perch, as the tide of corpses drifting into Lake Victoria slowly tapered off. But Rwanda's flood of the dead had been replaced by a living exodus. Refugees poured into neighboring Tanzania and Zaire as millions of villagers fled their homes: Tutsis escaping the massacres, and Hutus retreating before the victorious RPF army.

The refugees brought stories of horror that lent a personal touch to the tragedy—how a family survived for weeks at the bottom of a fifteen-foot pit latrine, hiding under raw sewage from the probing flashlights of the militia, and living on food dropped by a sympathetic neighbor. Or the entrepreneurs selling live honeybees to Tutsis, who stung their own faces repeatedly, hoping the swelling would make them resemble their broad-nosed Hutu counterparts. Years would pass before the country was reunited, and some estimates put the loss of life at well over a million people.

I pictured Tom on the hospital cot, whispering my name and struggling to raise his head. *This is only one death*, I thought bleakly, and the words followed me all the way home, a percussive echo for every heavy step: *one death, one death, one death.*

((((

I returned to Bwindi haunted by Tom's passing, and by the enormity of the situation in Rwanda. Somehow, the loss of a friend, however unrelated, brought an immediacy to the Rwandan crisis that had been missing in the months before. While Rwanda lay less than thirty miles away on a map, the difficulties of rural transportation stretched that landscape into something far larger. To most Ugandans, distance was an aspect of time: three hours to village X, two days to the capital. People rarely travelled without good cause; the process was expensive, inconvenient, and intensely uncomfortable. Most locals had never been farther than a day's walk in any direction, making life in a place like Bwindi seem deceptively isolated. My village neighbors viewed the Rwandan crisis as something remote and intangible. Although mortar fire was audible from the hilltops, everyone knew that the conflict was days away, in a strange place they'd never seen. We listened to the BBC or Radio Uganda to find out what was happening just down the road.

Arriving depressed and dusty from the journey, I found more bad news waiting for me at the park office. Distraught over the cancellation of his project at Mgahinga, German conservationist Klaus Jurgen-Sücker had hanged himself from the rafters of his Kisoro home. It was as if tragedy were seasonal, and this was the month for death.

"He was a very tall man, and it's such a small house," a high-ranking CARE official mused later. "He must have gone to a lot of trouble."

Too much trouble, in the eyes of his German employers. BRD sent an investigative team to Uganda, accusing CARE of conspiring with local poachers to plot Klaus's murder. Their case was high in passion, but had little basis in fact. Klaus had been visibly depressed for weeks, and tried several times to contact a priest in the days before his death. While he had enemies in the community, what murderer would have waited five years, only to kill him as he was literally packing his belongings to leave?

For months, the two opposing expatriate camps had lobbied for influence in Mgahinga. CARE and IGCP spoke with the power of American and English aid programs behind them, while BRD used the weight of the German embassy. Each side promised better management and greater financial support, and Uganda National Parks vacillated between them. But the murder accusations set off a new firestorm of controversy, resulting in high-level meetings, and diplomatic words between the ambassadors. Eventually, the Ugandan government terminated all BRD activities in the country.

During all of this, the park and the gorillas went with no support at all. While foreigners argued and jockeyed for the rights to future funding, Mgahinga's rangers worked for months without pay, food, or equipment. The common goal of preserving mountain gorillas became lost in the politics of conservation, a theme that carries into other aspects of international aid: relief agencies in Rwanda competing to serve the *most* refugees, at the *largest* camps; development groups fighting over big-budget projects; or whole governments struggling to expand their spheres of influence with no thought to the ramifications at a local level.

On the day of Klaus's funeral, I received good news from the world of gorillas, as if to counterbalance the recent waves of human tragedy.

"Mubare group has produced!" Phenny told me excitedly. That afternoon, the trackers had noticed one of Ruhondeza's six females cradling a tiny infant, still wet from the womb. It was the first new gorilla in over two years, a positive sign that any stress from habituation was no longer impacting their breeding success.

I put all thoughts of loss behind me the next morning as we neared Mubare group, climbing the steep, densely forested hill that bears their name. James Tibamanya, the lead tracker, gave me a cool smile and motioned the tourists to silence. Perpetually hungover, he usually lagged behind until the tracking became difficult, then took over the point

position, cap askew, and led us straight to the group. Today we heard
them feeding from several hundred yards away, and approached slowly
through the sunflecks and shade.

Ruhondeza lay like a heap of black earth in a bower of shrubs, pre-
dictably asleep while his family foraged around him. Three of the juveniles
climbed into a low, spreading *omwifa* and played king-of-the-mountain,
grappling, hooting, and pushing one another from the tallest branches. I
scanned the group, but saw no sign of the infant. Tibamanya tapped my
arm and pointed with his lips to a dark shape, deep in the undergrowth
behind the sleeping silverback. Through binoculars I watched the young
female, Mamakawere, facing away from me and preening something just
out of sight. Then she stood, and for an instant I saw her baby clearly, a
wide-eyed, pale-skinned newborn, clinging to the dark bulge of her stomach as if it were all the hope in the world.

CHAPTER XIV

CHANGES

———————▼———————

Year after year
on the monkey's face
a monkey's face

—Basho, Japanese poet, *17th century*
translated by Robert Hass

Karema hammered the ground with a resonant, double-handed slap and stalked towards us, stiff-legged, her calm brown eyes suddenly flinty with aggression.

"Slowly, slowly," Levi whispered to the tourists, a wide-eyed German couple in matching yellow rain suits. We inched backwards, regaining our fifteen-foot distance, and crouched down in a patch of dew-wet ferns.

Karema advanced to the small *omwifa* tree where we'd been standing. She reached up, snapped off a leafy branch and settled back against the narrow trunk, shaking the whole tree with her weight. Food in hand, her mood relaxed immediately. The whole incident was simply her way to make us step aside, the same communication technique that gorillas use to establish dominance and feeding privileges within the group. Still, any sign of hostility seemed out of character for Karema, the most gentle, even-tempered Katendegyere ape.

Levi glanced in my direction with an 'I-told-you-so' smile. He and the other guides had warned me of Karema's new-found surliness—they thought she might be pregnant. I watched carefully through binoculars, but her bulging midsection looked the same as any normal gorilla stomach. With a digestive system similar to our own, great apes lack rumens or other adaptations to process plants efficiently. They must take in huge quantities of vegetation to survive, and mountain gorillas spend more than a third of their waking hours foraging and feeding. Adult females can eat twenty pounds of fruit, leaves and bark in a single day, and silverbacks average twice as much. Their constantly-bloated stomachs make determining pregnancy a great challenge for researchers. Dian Fossey wrote of her shock and disbelief when Puck, the energetic young 'male' in one of her primary study groups, suddenly gave birth.

Karema fed calmly on the *omwifa's* fruit and leaves for close to thirty minutes. As I scrutinized her, I began to realize that Fossey's Puck-confusion might answer all of our questions. Whenever Karema stretched upwards to grab a branch or cluster of small fruit, I noticed surprisingly burly muscles rippling under the smooth black skin of her chest. Facially, she had always resembled Mugurusi and Kacupira, with long, narrow features that gave all of them a haughty, almost aristocratic appearance. But now, her whole head reminded me of the silverbacks, with a subtle pointiness that echoed their pronounced, and distinctively male, sagittal crests.

"I think we have made a mistake," I told Levi on the walk home. "Karema is a man."

He looked incredulous. A ranger-guide since the park was created, Levi Rwahamuhanda knew the gorillas as well as anyone. "No...is it?"

That night I pored over old photographs, picking out the slight, but noticeable progression of Karema's male features slowly taking over. The guides and trackers accepted this change grudgingly. We'd all come to think of Karema as a calm young female, helping balance the group's unusual, and aggressively masculine social dynamic. His sudden transformation skewed the sex ratio even further, bringing the number of mature males to six, all vying for the attention of a single female, Nyabutono. Her child, Kasigazi, remained a mystery, but its rambunctious play behavior had led the trackers to choose a name meaning "little *guy*."

In the field, young gorillas resemble one another too closely to recognize their sexes. Barring a physical examination, the most reliable method is simply to wait. When males reach adolescence in their ninth or tenth year, the differences gradually become obvious. Like Karema, they develop strong, hairless chests and a distinctive 'crest' along the upper sagittal area of the skull. Continuing to grow, they eventually double their sisters in size, and begin showing silvery-grey hairs on their lower back at age thirteen or fourteen. Contrary to public opinion and local myth, the 'silver back' is not an honor reserved for group leaders. All males grey as they age, their light saddle eventually spreading down onto the flanks and thighs. Females reach sexual maturity during this same period, settling the question by giving birth to their first child at age ten or eleven.

As if proving a point of machismo, Karema grew more belligerent in the months that followed. He started shadowing Katome on mock-charges, and we found him one day with a fresh head-wound, evidence of his first major intra-group scuffle. But through it all he seemed hesitant and uncomfortable in his new role, like a teenager reluctantly succumbing to peer pressure. Eventually, he settled back into old habits, leaving the charges and other histrionics to his more-excitable older siblings.

On one rare sunny day at the height of the October rains, he walked quietly towards us and lay down to rest, kicking his feet in the air and yawning, with an arm flung over his eyes to block out the light. The tourists snapped dozens of pictures before turning to the guide.

"Which gorilla is this?" someone whispered.

"Karema," he answered simply. "She is our newest male."

((((

Late afternoon sunlight angled golden over the forest as I sat on the porch, writing guide evaluations and sipping tea. I heard footsteps on the path and my new Peace Corps neighbor, Karen Archabald, stomped into the house, threw her notebook on the table, and settled into a chair.

"I don't understand why everyone here is so fascinated with my breasts!"

I smiled and offered her a cup of tea. "Did the women's group cop a feel again?"

"Yes," she laughed, exasperated. "I was talking, and someone tried to lift up my shirt."

I'd heard the same complaint from other female volunteers: Ugandan women seemed to have an insatiable curiosity about *muzungu* breasts. "I don't know what to tell you," I said helpfully. "The trackers don't seem too interested in mine."

"Shut up."

Karen arrived in Uganda only four short weeks after graduating from an Ivy League college in New England. She made the transition with amazing ease, aided by her great sense of humor and a riotous laugh that could bridge any cultural gap. When Karen found something funny, no one around her could help but join in the laughter

"This *Kareni*," the campground chairman told me after their first meeting. "She is jolly, jolly indeed."

Karen moved in to a small hut up the slope from my own, becoming my nearest full-time neighbor now that Dominico had sold his *shamba* to the park and moved farther into the village. She would often wander down in the evening for a bowl of beans and potatoes, and we quickly developed a camaraderie steeped in the humor of life as village *muzungus*.

Beyond the laughter, however, Karen's great asset as a volunteer was her innate empathy for the people she worked with. This compassion would draw her deep into the web of joy and tragedy that is life in rural Uganda. It equipped her well for the daunting task of filling John Dubois' shoes as Buhoma's community-development volunteer. Three years in the village had earned him legendary status as an organizer, motivator, and counselor for community projects. Luckily, Karen met John for several days of training in Buhoma, time for him to pass on some of his experience and site-specific knowledge.

"'Don't loan Benon any money,'" she quoted later.

"What?"

"That's all John told me," she said, throwing up her hands. "That's all I have to go on here."

"Well, it's good advice," I laughed. "He'd never pay you back."

"Yeah, but who the hell is Benon?!?"

To mark the end of his time in Uganda, John threw a three-goat party and invited the whole parish. Hundreds of people crowded into the old campground, eating, dancing, and drinking *tonto* in homage to the curly-haired man from Connecticut who had devoted a part of his life to their village.

"He is my son!" Dominico crowed, a sentiment echoed by the entire community. The park staff and scout troops mourned to see John go, posing for endless rounds of pictures in the fading gold light of his last Buhoma evening. The women's group composed a special song; the parish chief made a speech; and Phenny and Hope brought their newly-born daughter Amanda, giving the infant a final glimpse of her famous 'Uncle John.' Hope was still weak from the birth, but she and Phenny looked radiant, their pride as new parents overcoming the sadness of bidding John farewell.

John's goodbye party marked a turning point for me in Buhoma. During my second year, tourism in the forest increased dramatically and the tempo of village life accelerated to keep pace. From a fledgling park with two wardens and a borrowed motorcycle, Bwindi had grown into the major source of revenue for the entire National Park system. Six full-time wardens and an equal number of expatriate advisors struggled to coordinate construction projects, anti-poaching, revenue sharing, conservation education, and multiple use. IGCP extended its activities into Mgahinga, and Liz found herself tied up with meetings and workshops, sometimes visiting Buhoma less than once a month. I took over much of the project's on-site administration and spent every morning in the office, helping manage a chaotic visitor check-in process we came to know as "the tourist shuffle."

With Rwanda's civil war spilling over into Zaire, more and more tourists chose Bwindi for their gorilla tracking experience. Our permits sold out months in advance and we often had people waiting on stand-by, camped out in Buhoma for days at a time. Most visitors arrived well-informed and ready for a day in the forest, but with language barriers from Rukiga and English to Japanese or Dutch, the possibilities for miscommunication were endless. Common complaints ranged from "It's raining, I want a refund" to "The gorillas are nice, but what you really need is a gift shop." The guides laughed for weeks after one woman showed up in high heels and makeup, wondering where she could catch the mountain gorilla mini-bus. When we explained the rigours of tracking, she pretended to faint, swooning into the arms of a nearby ranger, then demanded her money back for reasons of health. But in terms of sheer drama, we spent our most exciting morning with a visiting Israeli general. He didn't have a permit, but demanded access to the gorillas on grounds of diplomatic immunity. His Ugandan escort, an army colonel with dark sunglasses and a beret, called park headquarters on the radio, threatening to hold our warden at gunpoint until we agreed to 'cooperate.'

The gorillas adjusted to their celebrity status with aplomb, calmly indifferent to the whir of cameras and video. Both Mubare and Katendegyere groups were fully habituated now, receiving their daily visitors with only an occasional charge, chest-beat, or other sign of stress. I accompanied the tourist groups at least twice a week, monitoring the gorillas, evaluating the staff, and finding new topics for my bi-weekly guide training sessions. While we covered everything from ornithology to first aid, it was my lectures on gorilla behavior that prompted the liveliest reaction from the guides.

"For gorillas, the forest is a restaurant," I told them one day, gathered under a shady tree near the office. "Their diet includes well over sixty different plants and trees, as well as fruit, grubs, ants, and even their own dung."

Everyone looked horrified, and stopped taking notes.

"It can't be!" Medad exclaimed in disgust and disbelief. "Not the dung. We have never seen it. The trackers have never seen it."

I assured him that Fossey and other researchers had observed the practice many times, suspecting that certain nutrients could only be absorbed on their second journey through the gorillas' inefficient stomach. But the guides were adamant: no self-respecting Bwindi ape would ever feed on its own excrement.

"Maybe for those Virunga gorillas," Phenny concluded. "You know they are dirty."

Combined with the in-forest work, shuffling tourists and training guides kept me busier than a *tonto* bar on market day. I sometimes found myself pining for the old days when Buhoma was still a sleepy, one-goat town. John and I laughed and reminisced about the past whenever I saw him in Kampala. He had stayed in the country for more than a month after his Buhoma departure, working on the one task he had yet to complete in Uganda. Through a generous donation of surplus equipment, he secured himself a temporary teaching post at Makerere University. The subject of his class? 'An Introduction to American Baseball.'

((((

Prunari knelt down and rummaged through the matted leaves and twigs of a large night nest. "Mutesi," he concluded, plucking out a short black hair and holding it up in the sunlight. He stood, brushed his knees, and handed the evidence to me: a wiry bristle, like horsetail, with the distinctive grey tip of a silverback.

"And here is Makale," said Charles, pulling all-black hairs from another nest several yards away. "From yesterday," he added. "The dung is old."

I sighed and checked my watch. Late afternoon sun angled through a broken sky, throwing cloud shadows across the forest like vast, drifting inkstains. From Hakanyasi, the landscape spread around us like features on a relief map: Bwindi's green ridges gave way to coffee trees and burned fields in Zaire, falling off sharply to the Ivi River valley and the great, flat-bottomed emptiness of the Rift. It was a beautiful place, but a long walk from Buhoma, and I'd already made the trackers work a double shift: visiting Katendegyere group with the tourists, and now, searching for the two renegade males.

As if to prove the gorilla world could change as fast as Buhoma, Makale and Mutesi had disappeared from Katendegyere group sometime the previous week. They went quietly, leaving no signs of a fight or confrontation with Mugurusi. The trackers simply noticed two missing night nests, and reported no sightings of either male. When several Zairian farmers complained at the park office about a pair of crop-raiding gorillas, we had little doubt that we'd found our runaways. The night nests on Hakanyasi confirmed it, but the trail was still a day old, and I'd hoped for a visual sighting. I decided to return in the morning and continue the search.

"*Ka Tugaruka*," I said to the trackers, and we headed home, bush-whacking downhill through the clearing's thick vegetation. Charles led the way, swinging his *panga* and leaning forward to push a path through the vines. We followed behind him, walking across a tangled mat of leaves and thick shrubs, suspended several inches above the ground. Suddenly, Charles shouted and leapt backward into me, knocking both of us down in a heap. I thought we'd stumbled into Makale until I heard Mishana and Prunari start laughing out loud.

"*Enjoka*," Charles admitted sheepishly, pointing with his *panga*. I peered into the greenery and glimpsed a flash of the reptile before it slipped away, something large and black like a forest cobra. With fourteen different varieties, including bush vipers, rhinoceros adders, spitting cobras, and other deadly poisonous species, Bwindi's snakes aroused more abject fear than any other rainforest inhabitants. And with good reason. Traditional remedies took time to prepare, and village healers held the proper mixture of herbs as a closely-guarded trade secret. The nearest Western serums were at Kisizi, over three hours away, if a vehicle were even available. Charles helped me up and laughed with relief while the others teased him, reenacting the encounter with exaggerated leaps and terrified shouts. But the levity died out as we set off again, and I noticed that no one offered to take Charles's place in the lead.

I arrived at the office early the next morning and found the trackers gathered around the cookpot out back, drinking hot mugs of corn meal porridge.

"*Erizooba, Ninronda Makale na Mutesi.*" I told them I wanted to return to Hakanyasi and spend the whole day tracking. Immediately, their eyes fell from mine and they started shifting uncomfortably. "Two of you have to go with the tourists," I went on. "But I need someone to help me find Makale."

They hated this. None of the staff wanted to get anywhere near the gorillas with fewer than four people.

"Makale will count," Medad used to say, squinting and nodding his head like a gorilla doing arithmetic, "and if you are too few, then he knows he can charge!"

Finally, Mishana volunteered to join me, stepping forward reluctantly, as if accepting latrine duty in a Siberian gulag. We climbed together through the overgrown tea fields behind my house, past the level plot of land where Dominico's family compound had stood. Overgrown yam plants grew up through the thatch of the fallen cook hut, broad leaves shining in the porcelain light of a cloudy Bwindi morning.

Crossing into the forest, we startled two palm nut vultures from their roost in a leafless canopy snag. They flapped briefly and drifted out over the valley, their wings a geometric study in black and white against the deep greens of the opposite ridge. We took the Muzabajiro trail and hiked towards distant Hakanyasi, hoping to find Makale and Mutesi before they shifted to another area. Lone males or sibling pairs often shadowed the movements of their old group mates, but their ranges expanded quickly as they gained more confidence on their own.

Emigration plays a vital social and biological role in gorilla populations, helping maintain a constant flow of individuals and genes between the various families. Most males leave their natal group when they reach sexual maturity, travelling on their own or in sibling pairs like Makale and Mutesi. As they gain in strength and experience, they begin seeking out other groups and challenging the lead silverbacks. These interactions involve dramatic displays of charging and chestbeats, occasionally resulting in violent physical conflict. In rare cases, the intruding male defeats the resident silverback and takes over his role of leadership. More commonly, females use the conflict as a chance to 'shop around' for a new mate. A younger, low-ranking individual might choose to switch allegiance and follow the challenger, with whom she'll have higher status and a better chance of successfully rearing her children. Most females make several inter-group transfers during their lifetime, helping form new families, or carrying their genes to other, well-established groups.

With Mugurusi presiding over five growing bachelors and only one female, Katendegyere group had been long overdue for a change. In the months before their departure, Makale and Mutesi had spent more and more time together, moving on the periphery of the group and harassing the younger males. Mutesi was the larger and stronger of the two, but Makale seemed to dominate his older brother, leading the way whenever they charged. I once watched them displace Karema for no apparent reason, chasing him from a prime feeding area with repeated lunges and screams, only to settle down and sleep as soon as the young blackback abandoned his meal. Aggression within the group decreased noticeably after they left, the way unruliness might disappear from a high school gym class after suspending the chief trouble makers. But I found myself missing

Makale's surly antics, and I was glad when the warden-in-charge asked me to keep tabs on him from time to time.

As Mishana and I neared the top of Hakanyasi, we crossed a swath of flattened vegetation, scattered with newly-peeled stems. Mishana picked up a broken twig and rubbed it between his fingers.

"*Erizooba*," he concluded: 'today.' We had stumbled into a fresh trail.

Their path was obvious and I took the lead, following the bent leaves and trampled shrubs through a wide clearing near the border. Just as I began to think my tracking skills were improving, the trail led into a copse of trees where the understory opened up into a mossy swale. Three different trails crossed in from the left and I found a day nest, half-chewed *omwifa* fruits, broken tree limbs, and four piles of fresh dung. Several other trails led out of the trees—one uphill, two downhill, and one in the direction we'd just come from. I turned to Mishana and threw up my hands.

"Mubare," he said, smiling at my confusion. Apparently, Makale and Mutesi had crossed Mubare group's trail from the previous afternoon. Mishana walked quickly over the mossy ground and poked at a lone bolus of dung with his boot. Then he turned towards the uphill trail and motioned for me to follow.

Twenty minutes later, we heard branches snapping on the hillside above us, and we climbed more slowly, beginning to cough a greeting. Just then, the sound of a chestbeat drifted up from somewhere south of us on the ridge.

"Ruhondeza," Mishana whispered, and we paused.

There was no response from above. We continued on to the place where Makale and Mutesi had just been feeding, and there the trail took a sudden sharp turn, leading directly away from the sound of Ruhondeza's chestbeat. Apparently, our brave bachelors weren't quite ready for a confrontation with Mubare group. Ruhondeza's far-reaching note of warning had served its purpose, establishing his presence in the area, and avoiding

the chance of an unplanned encounter. It took us another forty minutes to
catch up with the wary males.

We found them near the remains of a fallen tree. Makale was feeding in
a dense thicket below us, while Mutesi searched the rotten log for termites. He squatted with his head out of sight beneath the trunk, and his
grey rump raised up awkwardly in the air. A fine mist had begun falling,
and I watched his huge, leathery fingers scrabbling for a hold on the slippery wood. He rocked the log violently back and forth; from my angle, he
appeared to be methodically picking it up and dropping it on his face. I
inched closer to get a better view.

"Ssst," Mishana caught my attention, and shook his head, gesturing
down the hill towards Makale.

Oh please, Makale can't count, I thought, and shifted closer.

Immediately, the shrubs below erupted with a screaming roar and
Makale charged towards me, lunging up through the chest-high bracken
with his teeth bared. I started a submissive crouch, but he stopped short
several meters away, looking suddenly disinterested, and spun around to
stomp back to his feeding ground. Mutesi hadn't even glanced up from the
termites, and the whole encounter seemed perfunctory, as if Makale were
simply trying to keep up appearances.

I sat down as the mist turned into rain, and pulled out my field notebook: "*11:35—Mak. charges to within 4 meters. Retreats immediately.
Embarrassed because we saw him run from Ruhondeza?*" The pungent, nerv-
ous-sweat smell of Makale's charge dissipated quickly in the wet air, and
the only sound came from Mutesi, gnawing quietly at the deadwood.
Finally, he moved off to find shelter, and the clouds settled lower, closing
the gap between earth and sky, and shrouding our green hilltop in layers of
impenetrable whiteness.

Chapter XV

Lessons

▼

*"Four thousand feet above my inn, between two mountain
peaks, I always felt as if I were on another planet...
The automobiles, racing with arrogant importance over the
roads of Africa, looked like flimsy toys through my binoculars.
Their speed seemed futile, their eagerness absurd. '
Why all this hurry?' I wondered."*

—Walter Baumgartel
Up Among the Gorillas, 1976

"O.K., Barigye, take it easy," I counseled, but the car continued to pick up speed, racing along the road-edge in first gear. "*Mpora, mpora, Ssebo*," I tried calmly, remembering that his English wasn't strong. But one glance at my pupil told me that Barigye wasn't hearing anything. He clenched the wheel in a death grip, eyes wide and glazed, teeth bared in a paralyzed leer.

Suddenly, we veered off the roadway, flattening a wide swathe of shrubs before us. "Jesus! Philman, tell him to stop!" I yelled, but Philman and Ephraim were just trying to hold on in the back seat as we bounced and jarred across the field, tall elephant grass beating frantically against the windshield. Barigye swerved again and we re-crossed the road at an angle, speeding towards the spring where a group of kids were lined up for water.

I saw their mouths open to scream as they darted out of the way, seconds before we finally lurched to a halt.

In the sudden calm, Ephraim and Philman started laughing hysterically and I breathed a sigh of relief, but Barigye remained silent. He looked dazed, still gripping the steering wheel, and slowly blinking the sweat from his eyes. I reached over and turned off the engine. "I think we've had enough driving lessons for today."

Barigye nodded mutely and hopped out of the car. On his first day with the park's tourism staff, he'd already been charged by Makale, and now his driving lesson had turned into a ride of terror. The following week, he gratefully accepted a transfer back to ranger duty, and the relative peace of armed anti-poaching patrols.

With Liz away for two months on home leave, I had use of IGCP's Maruti, a cheap, two-door Suzuki clone made in India. "I call it P. O. S.," she told me, handing over the keys: "Piece Of Shit."

The car rattled like a cookie tin filled with nails, but it lasted through years of backcountry use—overloaded market runs, rough roads, watery gasoline, and now, Buhoma's first official driving school.

"One lesson, Tour," Ephraim pleaded as I drove back to the house, but I shook my head no. Everyone longed for a turn at the wheel, but I wasn't sure the Maruti or the teacher would survive another trip like Barigye's.

"Maybe tomorrow," I told him.

"Then at least we should wash it," Philman put in. "You see the vehicle is very dirty from driving off the road."

I gave in, and we stopped by the creek, drawing our usual crowd of spectators. Although tourism had made vehicles a common sight in Buhoma, many people still regarded them with wary fascination. Other foreign technologies, like a Walkman or short-wave radio, aroused the same response: a sense of wonder that Western observers would describe as childlike. I never got over the unease of watching my friends marvel at things I took for granted. But at the same time I felt ashamed and saddened that my own culture equates amazement with immaturity. Where Africans view wonder as a natural reaction to the unfamiliar, we consider it valueless—the antithesis of knowledge, a response of youth and inexperience.

I helped Philman heave basins of water over the car, while Ephraim carefully scrubbed mud from the hubcaps. It was futile, of course. A fresh layer of dust and grime would splatter up the sides of the vehicle before we'd driven a hundred yards, but that was a part of their plan. After all, a perpetually dirty vehicle always needed a trip to the car wash, and maybe there would be time for a lesson along the way.

The next morning, I woke to a quiet knock at the door.

"Tour..." someone called softly. "*Kodi, kodi.*"

Dim, pre-dawn light crept in around the shutters as I stumbled out of bed and looked for a pair of pants. My house ran on an open door policy, and visitors stopped by periodically throughout the day, but they usually waited for the damn sun to rise.

"*Agandi*, Gongo," I said to the trim, grey-haired man on the lawn as I stepped out into the half-light. It was Phenny's father, Tibesigwa, a respected elder in the village, and the newly-appointed manager of the community campground. He returned my greeting, shifting his weight from foot to foot and digging one toe into the ground, as if reluctant to disturb me so early. Etiquette required a pause before coming to the point, and I watched his intelligent, bearded face as we chatted, trying to discern some hint of the morning's problem. Tibesigwa shared Phenny's impish sense of humor, but there was nothing funny behind today's visit.

"I hope you can help my granddaughter with transport," he said finally.

"What's wrong, Gongo?"

"She is producing too soon," he explained with an earnest look, and I saw the worry in his dark eyes. "There is a problem with the pregnancy."

"What about Proscovia?" I asked, referring to the nurse from Buhoma's tiny health clinic. "Is she there?"

"Yes, yes. She says the girl needs Kisizi."

Kisizi Hospital, the nearest medical facility of any size, lay over three hours away on bad roads. There were no telephones or ambulances, of course, and P. O. S. was the only vehicle in Buhoma. "I'm coming, Gongo," I told him. "Let me get ready, and stop by the office first."

"*Webale, Ssebo*," he shook my hand and turned to hurry home, calling over his shoulder: "You can pick her from the path by Buhoma church. Phenny will show you."

Early sunlight brushed against the hilltops as I drove towards the office, brightening the stark line where forest greens gave way to the cultivated shades of millet and young cassava. I found Phenny waiting by the roadside. He gave me a mock salute as I pulled up, and jumped into the passenger seat.

"*Tugyende hamwe, Ssebo*," he offered: 'we will go together.' I noticed that his uniform shirt looked neat and freshly pressed. With cheap laundry soap and a charcoal iron provided by the park, the guides always managed to dress sharply, while my own clothes—stained khakis and a t-shirt—looked like, well, like I lived in a jungle.

"All the way to Kisizi?" I asked him.

"Sure. I need to pay Hope's bill," he said. "And pick up more medicines." Sudden concern swept the smile from his face. Ever since giving birth to Amanda, Hope had suffered from stomach problems and chronic fatigue. Two trips to the hospital had failed to provide a cure, and her one-month maternity break had turned into an indefinite leave of absence. She remained cheerful, staying home and managing the growing restaurant business at the H & P Canteen. But any lingering illness was cause for concern, and Phenny's usual good humor had given way lately to frequent periods of brooding depression.

"How is Hope?" I asked, a question that had become as familiar as his inevitable answer.

"She is improving somehow, but still weak."

We drove in silence to the church, where a large crowd had gathered, surrounding the pregnant woman and her family. They lifted her gently into the back of the vehicle, tilting up one of the bench seats so she could lie flat, with her head resting forward. Her mother and three brothers crowded in beside her, burdened with bundles of food for a long hospital stay. I waved to Mr. Gongo, who came to the window, frowning, every inch the solemn patriarch. He exchanged words with the people in back, then turned to me.

"They have money for petrol," he said with approval. "One jerrycan at least. You can pick it from Butagota."

"It's no problem, Gongo," I told him, and eased the car into gear. The crowd stood back to watch us go, a jumbled knot of bright clothes and dark skin receding in my rear view mirror, with a hundred eyes staring after. We rattled towards Butagota, and I looked down at our young passenger. She was dressed as if going to market, her head wrapped in a blue kerchief, and dress arranged neatly over the bulge of her stomach. But she rode in obvious pain, wincing at every bump, with a fine sheen of sweat glistening on her smooth forehead. I stopped the car frequently, and her family raised her up, rearranging the cushions of bundled clothes, and making her drink from a calabash of sweet porridge.

"*Webale, Ssebo,*" she whispered once, 'thank you,' her round eyes gazing up at me, glazed with pain and resignation. It was an image that would stay with me always: laying my hand on her head in comfort as I braked and swerved to avoid the potholes, as if there were anyplace else to drive.

The sun rose hot in a dry season haze as we sped through Kigezi's hill country, raising clouds of dust behind us like a cavalry charge. I turned on some music to ease the waves of tension and worry that rose up from the back of the car. Phenny grinned, and sorted through the cassettes, but spent most of the ride staring out the window, lost in his own thoughts.

At the village of Kanungu we turned east on a narrow dirt track, climbing again, and then descending towards the broad valley where Kisizi lay tucked in a river bend, just downstream from the waterfall that bears its name. The town stretched along a quarter-mile of dusty roadway, a small collection of shops and stalls that served as a trading center for local farmers, supplying the busy hospital and its constant stream of customers.

I drove slowly through the market to the hospital itself, several large brick buildings and a neatly tended lawn that dominated the village center. People milled everywhere—nurses, uniformed orderlies, and throngs of visitors, walking together in small groups, or waiting beneath the huge eucalyptus and jacaranda trees that shaded the compound. I dropped my passengers at the office and went to park, while Phenny settled Hope's bill and helped his ailing young relative navigate the check-in process. When I returned, they had already carried her off to the maternity ward.

"Let us go see the waterfall," Phenny said, glancing at his wristwatch. "Afterwards they should know if she needs to stay on."

"Did they say what is wrong?"

"It's probably nothing," he said dismissively. "People in the village don't know anything, so they get excited."

He turned away and I followed him towards the sound of the waterfall, but I wasn't so quick to disregard the woman's problem. With so many tropical maladies unknown to Western science, people in Buhoma couldn't always put a recognizable name to their illness, but they knew when they were sick.

After I had learned to bandage the trackers' occasional *panga* wounds, they began sending me their friends and neighbors with more complicated ailments. Any sickness accompanied by a fever was called 'malaria,' while head colds and congestion were known collectively as 'flu.' "I am weak," my patients would complain, pointing to the afflicted region. Sometimes, their foreheads or aching joints would be cross-hatched with tiny scars, where they'd cut the skin repeatedly with a sharp knife or razor blade. A nurse in Kabale explained this method as "letting the millet out," an attempt to bleed away painful symptoms. She told me that it helped some people, but was dangerous in infants, whose parents often inflicted deep chest wounds to rid them of croup cough, or cut out unformed baby teeth as a cure for diarrhea. But many local remedies worked far better than their pharmaceutical counterparts: tree bark teas for stomach worms, an herbal poultice for snake bites, or certain crushed leaves that instantly staunched the flow of blood. In my small 'practice,' I relied heavily on the placebo effect, doling out aspirin, Band-Aids and vitamin C, and referring everyone to the nearest clinic.

Sometimes, a malady came along that baffled local and Western traditions alike. Ephraim Akampurira had recently missed over a month of work, sick in bed with a terrible case of 'malaria.' Philman told me he'd never seen an affliction quite like it, but this really came as no surprise. If anyone would catch a strange illness, it would be Ephraim, who proved himself time and again as the most hapless, accident prone man in the village. He injured himself constantly—sprained ankles and snapped collar bones from playing soccer, skinned knees and gravel-road abrasions from bicycle wrecks, tumbling out of moving vehicles, or simply tripping and falling over. "I have broken my face," he announced after one such mishap, arriving at work with swollen cheeks and a wildly bloodshot eye.

I'd been sending his salary home to the village, along with packets of aspirin and vitamins. "He is improving," Philman announced finally, and two days later Ephraim came back to work. He smiled and shook my hand warmly, but seemed unusually quiet when I invited him in for tea.

"You are welcome back!" I told him.

"Mmm," he mumbled, and the conversation didn't go much further.

The next morning he arrived with Philman, who took me aside while Ephraim looked on, earnest and silent.

"Ephraim is wondering whether he can continue working here," Philman said seriously.

"Well, why not?" I asked, a little surprised. "He wants to go?"

"No. It's just that he's forgotten how to talk."

"You mean he's forgotten English?" I clarified, thinking his vocabulary might have grown rusty from a month of disuse.

"No, no. He can't even speak Rukiga." I looked at Ephraim, who shrugged his assent. "He understands," Philman pointed out, "but he can't find any words. It's from the malaria."

"Ah." I paused and glanced at their worried faces, making sure this wasn't some kind of elaborate joke. "Well, he knows the work here," I said reassuringly. "Of course he can stay."

They grinned and we all shook hands. In the weeks that followed, Ephraim's speech slowly improved, but he never regained his former fluency. I asked doctors from Kabale, Kampala, and even the States, but no one recognized the symptoms of 'language fever.' It made me appreciate the random disease risk of life in Buhoma, and I started saving a few vitamins for myself.

Phenny led me through a grove of tall eucalyptus behind the hospital. Originally imported to help drain Uganda's swamps and lowlands for farming, the water-hungry trees were now a popular, fast-growing source of firewood and timber. Their voracious root systems couldn't drain a river, however, and the path grew muddy as we approached Kisizi falls. Cool mist surrounded us as the cataract came into view, a torrent of whitewater plummeting more than a hundred feet from the cliff face into a turbulent, rocky pool. Green moss and ferns clung to the sheer walls of the grotto, and the trail stopped at the water's edge, under the squat, glistening leaves of a wild banana.

"Before you whites came, that girl would have died here," Phenny shouted over the roar of the falls.

"You mean without the hospital?"

"No," he shook his head. "This place. The waterfall. This is where the locals used to judge unmarried mothers. They threw them from the top there." He pointed to a ledge that jutted out from the top of the cliff, overhanging the waterfall's precipitous drop. "No one could survive."

The practice, I learned, disappeared around the turn of the century, but inspired Christian missionaries to choose this otherwise out-of-the-way village as the site for a rural hospital.

When we returned to the maternity ward, a disinterested nurse told us that we could go home.

"She will spend the night here at least," the woman said, shuffling through some papers. "Maybe two or three."

I asked for a diagnosis, and she grew annoyed. "The doctor said she will stay. So she stays."

Phenny pulled at my arm and rolled his eyes as if to say 'don't waste your time.' He knew from experience that battling Kisizi's bureaucracy was futile. Hope had spent a full week at the hospital, only to be sent home with a bill, a handful of medicines, and no explanation of her symptoms. We left without seeing his relatives again, and stopped in the market for lunch: balls of sweet fried dough and a bundle of *kabaragara*, the finger-sized yellow bananas.

Driving home, the road climbed along the edge of a steep valley. Phenny peered out of the window and pointed down. "My mother's family is from this place. Can we stop?" he asked. "I need to get one small cousin."

"What?"

"I need a kid to help with Amanda while Hope is sick," he explained.

The ultimate babysitting solution, I thought, and tried to imagine dropping in unannounced on my cousins' family back home, and kidnapping one of their children. "Sure," I told him as I pulled over. "But be quick, Gongo. I want to get back before dark."

He grinned and thrust a bunch of *kabaragara* towards me. "You take these bananas," he said as he leapt out of the car. "I will return before you finish them!"

The sun was sinking towards Zaire and I sat surrounded by tiny peels when Phenny finally came back, sweating from the steep climb, with his shirt up over his head.

"Where's the kid?" I asked, looking behind him on the trail.

"She will come later," he replied and hopped into the car.

"By foot? It's too far." The distance to Buhoma was more than forty miles.

"It's O.K. She has these Ugandan vehicles," Phenny said, and held up one of his bright plastic sandals.

I shook my head as we rattled away up the road, laughing together in the reddening light.

((((

I gripped the wheel like a tracker on his first driving lesson as the car slid sideways in the sand, then straightened out, accelerating towards the bottom of a rocky gulch.

"Ahh, Tour!" Phenny crowed as we hammered across the gully. "You really know driving!" He laughed as I punched it into first and began climbing a steep wall of bedrock and boulders. We hit a patch of gravel and the front tires spun, pelting the bottom of the car with loose stones. "Kareni could never drive this one. Not even Dr. Liz!"

"They could, Phenny." I said, gritting my teeth. "The point is that they wouldn't." *They're too smart*, I thought. *How the hell did I get myself into this?* Finally, the "road" levelled out into a dusty footpath, and we bumped along, flattening shrubs and scraping through narrow tunnels of elephant grass.

"They used this track for smuggling," Phenny told me. "Mostly in Amin's time."

Apparently, no one had driven here since. We were somewhere north and east of Butagota, descending gradually through a parched landscape of small farms near the Zairian border. Astonished people leapt out of the way as we approached, and children sprinted along behind the car, waving and shouting. Even the goats looked surprised to see us. We passed a tiny churchyard, where two soccer teams chased a ball of tightly-wound banana fibers. The players stopped their game and stared after us, shaking their heads as if we were just as crazy and out of place as I felt.

"Don't mind," Phenny said to reassure me. "The house is close."

But we rattled on for another hour of hot sun and dust. In a few short weeks, it seemed that my role in Buhoma had switched from Peace Corps volunteer to taxi driver, with constant trips to the market and medical clinics. Where we all used to 'suffer together' without a vehicle, I now found myself honor-bound to pick up the rangers' monthly *posho* supply, or haul sacks of cement for the campground. I began to dread people's transportation requests, particularly when the results were beyond my control. Earlier that week, funeral drums had announced services for Mr. Gongo's granddaughter, who died at Kisizi along with her unborn child, in spite of their ride to the hospital.

Today's journey was another ambulance run, to the home of a respected local healer where Hope had gone for treatment. Her coughing and weakness persisted in spite of the medicine from Kisizi, and little Amanda had taken ill as well, with painful sores in her mouth that prevented proper nursing.

"It is a curse," Phenny explained to me that morning. "People in the village are jealous of Hope's success with the restaurant, the park job, even helping the campground. Someone has cursed her for it."

We had stopped along the road to pick oranges, the bitter green variety that Hope liked. "They are too sour," Phenny said, making a face, "but she always wants them." He walked under the tree, shaking the branches with a long stick until enough fruit rained down to fill a sack. I asked him about the curse and he told me how people prayed to their ancestors for help against an enemy. "Local religions are still here," he said. "A witch doctor can call up spirits from your family to make trouble for the living. The people at Kisizi Hospital can't help with this. Only local treatments."

We came to a fork in the path and Phenny hopped out to ask directions. He was eager to reach the herbalist and bring his small family back to

Buhoma. But we'd brought the oranges, and a basket filled with eggs, bread, and tomatoes, just in case Hope needed to stay on for more treatment.

"It's just ahead," he said, and we bumped along the narrow track to a tidy-looking brick house, half-hidden behind a tall, evergreen hedge. I pulled into the shade to park, and waited by the car while Phenny disappeared into the compound. A group of children appeared, staring at me wide-eyed as I stretched my legs after the long drive. I made faces at them to pass the time, then noticed a skinny old woman shuffling towards the car, smiling broadly as if she knew me.

Who's this, I thought briefly, and then recognized her: *My God, Hope.* She walked as if exhausted by the effort, a faded dress hanging from her gaunt frame like scarecrow's clothes. Only her eyes were recognizable. Her once-round features had gone thin and ancient, cheek bones protruding like sculpture.

"*Agandi*, Tour," she said, and thanked me for coming. "I can't believe you drove all the way here."

I took her hand, and greeted her automatically, hoping a smile and familiar words would hide my shock. All I could think of was Tom Ntale, the last person I'd seen so wrecked by disease.

"The baby is fine now," she told me as Phenny came out of the house with Amanda. "You can take her home, but I think I'll remain here another week. I'm still improving."

"Of course," I said, turning to unload food from the car. "We'll come back for you, Hope."

I walked her back to the house and met the herbalist, a huge, regal woman who laughed and laughed when she saw me. "*Muzungu*," she muttered, shaking her head in disbelief as she ushered Hope back inside. A curtain of beads rattled across the doorway behind them, and I could still hear her chuckling as I walked away.

The drive home was long and quiet. Phenny sat with Amanda cradled in his lap, and I stared numbly ahead, sure that I would never see Hope alive again. But ten days later the old herbal healer proved me wrong, and

Hope came back to Buhoma, feeling stronger than she had in months. Her appetite returned and she put on weight, and even came back to work part time, her calm presence in the office a brief, but reassuring sign that the curse had been lifted.

———

Chapter XVI

Visitors

———————————▼———————————

*"One morning I was sitting near William a short way up the
mountain above camp when a boat arrived with some visitors
from Kigoma...I should of course have gone down to say hello,
but I had become so attuned to William that I almost felt myself
the chimps' instinctive distrust of strangers. When William moved
down toward the tents, I followed him; when he sat in the bushes
opposite my camp, I sat beside him. Together we watched the visitors...I wondered what they would have thought if they had
known I was sitting there with William, peering at them as
though they had been alien creatures from an unknown world."*

—Jane Goodall
In The Shadow of Man, 1971

I placed my feet between the roots and clumps of foliage, inching silently
up the muddy path. Overhead, soft grunts murmured through the canopy
like a distant conversation, and I heard fruit pith dropping through the
leaves all around me. I headed towards the closest gap in the forest, where
a wind-blown tree had opened narrow views into the tangle of branches
above. Looking up, furtive movements drew my attention to the topmost
layer of green, a sprawling limb of the giant fig tree I could see from my

back yard. Between the emerald leaves I glimpsed a face peering back at me, shiny and black, its wide mouth frozen in a Cheshire leer.

I stood absolutely still, bracing myself for an ear-splitting scream, but the chimp held its peace. It stared at me for a calculating moment, then continued feeding with exaggerated nonchalance, as if deciding that one lone *muzungu* wasn't much of a threat after all. *Progress*, I thought, and fumbled for my binoculars.

On closer inspection I recognized the chimp–a large male I'd come to think of as perhaps the most dominant in the area. With a flat, protruding brow and a bald patch stretching back over the top of his head, he looked like an old monk wrapped in thick, hairy robes. I watched as he leaned out precariously and snatched a cluster of sun-ripe figs from the highest reaches of the tree. He held them close against his chest and squatted on the branch, thumbing the green, plum-sized fruits into his mouth and chewing sloppily. Soon, his lower lip bulged with pith as he sucked the juice between his teeth. Known as a 'wadge,' this method gives feeding chimps a perpetual, clownish smile, like children playing ape with slices of orange.

Over the past two weeks I'd grown used to the sight, and to the chimps' excited hoots and shouting as they arrived at the tree each morning. Living in scattered sub-groups throughout their home territory, chimpanzee troops keep in contact with a range of loud vocalizations. In this case, the message of their shrieks was unmistakable: "Fruit! Ya-hoo!! *FRUIT!!*" These figs, a favorite food source, drew chimps from a wide area and made them unusually bold. Instead of taking flight at the first sight or sound of people, they risked feeding in open view of my house, and even let me approach to the base of the tree. My job description didn't include habituating Bwindi's chimp population, but I couldn't resist the opportunity to observe the apes up close.

Theoretically, chimpanzees and gorillas diverged from a common ancestor at roughly the same time they split from humanity, between six and ten million years ago. We all share more than ninety-eight percent of the same genes, but that small remainder accounts for significant differences between the species. Behavioral scientists debate the merits of each, noting the chimpanzees' complex and dynamic social structure, their problem-solving skills, and a talent for tool making. In contrast, gorillas live in comparatively stable families, and show a great capacity for imagination, memory, and patience. Any gathering of primatologists divides quickly into the chimp people and the gorilla people. The chimp-lovers dismiss gorillas as indolent, boring apes that soil their own beds every night, while the gorilla fans accuse chimpanzees of guile and violence, with their organized monkey hunts and attacks on neighboring troops. Where chimps are excitable, gorillas tend towards calmness, but individual personalities vary dramatically in both species, as much as they do in their human observers.

Leaves rustled above and a younger chimp appeared, scooting out the branch towards the male. It had the pale bronze skin and white-tufted rump of a juvenile, and held its face in a loose-lipped grin, as if contemplating something funny. The joke turned sour when it spied me gazing up from below, and it screamed suddenly in fear and surprise. The noise, an ascending soprano whoop, brought a loud response from the crown of the tree, and the juvenile leapt back into the greenery, followed by the old male. Several nights earlier, the same shrieks had transfixed me on my way to the latrine, when light from my lantern disturbed a group sleeping in the yard. In the darkness, their voices had been eerily disembodied, but now I saw chimps everywhere, dropping out of the tree like paratroopers. They fled in complete disarray, a deluge of tumbling shadows silhouetted briefly against the understory's dim mosaic of vines, saplings, and specks of daylight.

I held my breath as they melted away into the forest, then eased closer to the tree, sensing an unnatural silence overhead. Sometimes, when chimps try to be quiet, they create a conspicuous absence of sound; their lack of activity is almost a noise in itself. After a few minutes, fruit pith began dropping again, then the steady patter of urine. Five chimps remained in the tree, feeding on the bounty of figs left behind by their more nervous comrades. I found a comfortable seat out of range and settled in to watch and listen.

An iridescent flash caught my eye, and I followed the flight of an olive-bellied sunbird to its nest. The tiny husk of dried grass hung suspended like woven fruit from a natural marling of twigs and blue-flowered vines. I watched the bird enter briefly and take wing again, three times in the span of five minutes. Like the humming birds of temperate climes, African sunbirds are nectar feeders, with long curving beaks and a lightning-fast metabolism. They rarely perch for longer than a few seconds, constantly flitting through the undergrowth in an endless, frantic search for blossoming plants. Thinking of the recent pace of my life in Buhoma, I could relate.

With tourism booming and the end of my Peace Corps service in sight, I dashed from one project to the next, struggling to wrap things up. Both Katendegyere and Mubare groups were fully booked now, and extra people still turned up at the office every morning, adding their names to a stand-by list that sometimes grew to forty or more. Travel agents in Kampala competed hotly for our limited tracking permits, as film crews, journalists, backpackers and tour groups from Europe, Australia, Japan, Canada and the States all vied for a chance to see the gorillas. The park planned to habituate a third and final gorilla group for tourism, so I hired extra trackers and began training new guides, swelling the staff at Buhoma to more than thirty people. The new Warden of Tourism helped manage our daily tourist shuffle, and I spent more time working at home, compiling a guide training handbook, writing trail pamphlets and fact sheets, coordinating supplies, schedules and accounts, and fixing cups of milky tea for a seemingly endless stream of visitors.

Even now I could hear voices at my house, a hundred yards away across the ravine. I found a view through the intervening foliage and saw Ephraim, patiently pointing out the chimp tree to a group of curious tourists. Every day, dozens of people climbed the hill to my place for a visit, from tour leaders to park staff, travellers, friends from the village, or strangers come to meet the Buhoma *muzungu*. "Welcome to the tea house," Liz used to say on busy days, but now, our 'tea shop' had more business than a mall food court at the height of the Christmas rush.

With so much work left to do, and the promise of new gorillas to habituate, Liz, Karen, and the wardens had all encouraged me to stay on in Bwindi. In certain cases, the Peace Corps allowed volunteers to extend for a third year, the way John DuBois had stayed to finish Buhoma's Community Campground project. Sitting in a patch of sun-flecked shade beneath the fig tree, with chimpanzees feeding calmly above, the prospect of another year looked incredibly appealing. But another reality awaited me outside the forest–the subtle, but mounting strains of life as a village *muzungu*.

I joined the Peace Corps hoping to find a sense of place in a foreign community, and felt I'd reached that point in Buhoma. But my place was a little strange. Outside the close circle of my friends and co-workers, people still regarded me as a kind of bizarre addition to the village landscape. When explorer Richard Francis Burton first penetrated the East African hinterlands, he found that a host of mythical expectations preceded him:

> *"They (muzungus) had one eye each and four arms; they were full of 'knowledge,' which in these lands meant magic; they caused rain to fall in advance and left droughts in their rear; they cooked watermelons and threw away the seeds, thereby generating small-pox; they heated and hardened milk, thus breeding a murrain among cattle; and their wire, cloth, and beads caused a variety of misfortunes..."*

I hoped that I hadn't advanced any of these theories, but knew I would always be an outsider to most people, the object of curiosity and constant scrutiny. Peace Corps literature described this effect as "life in the fish-bowl," and it led to a sometimes-comical lack of privacy. I often carried on conversations with people who stood outside the shower stall while I bathed, chatting about crops, weather, or the health of their families.

Over time, this unyielding attention took its toll on any visitor. In Burton's early journals, he reacts with bemusement:

> *"...every settlement turned out its swarm of gazers, men and women, boys and girls, some of whom would follow us for miles with explosions of 'Hi-i-i!,' screams of laughter and cries of excitement."*

Halfway into the two-year march, however, he sounds on the verge of a breakdown:

> *"We felt like baited bears: we were mobbed in a moment, and scrutinized from every point of view... Their eyes, 'glaring lightning-like out of their heads,' as old Homer hath it, seemed to devour us..."*

I hadn't noticed any Homeric lightning bolts yet, but there were certainly days where I longed for anonymity.

I had also begun reaching the limits that culture placed on my relationships with friends. While I'd formed close bonds with Hope and Phenny, William, Alfred, Ephraim, and many others, the sheer disparity between our backgrounds often prevented a deeper connection. We worked and laughed together every day, but found our conversations at an impasse where cultures diverged. At these times the friendships became instinctive: we still felt close, knowing how much better we'd communicate if only we'd grown up in the same world.

"Ah, Tour, we will be missing you," Alfred said sadly, when I told him I'd be leaving at the end of two years. "You should get a wife and one small banana *shamba* in the village, and just stay here."

"No, no. Not a *shamba*," I replied. "I will open a new *tonto* bar, with discos and video shows every night!"

I joked to cover my discomfort about the decision. It was a Ugandan trait, levity as counterweight to sorrow. But in spite of my ambivalence, I knew that departure was the right choice. I'd met several volunteers and expatriates who tainted their whole experience by staying abroad too long, by letting life in the fishbowl turn cultural stress to bitterness. I wanted to leave Uganda on a positive note, when it would still be hard to go.

((((

A chestbeat rang out through the mist, haunting and urgent, like a three-tone message drummed on hollow wood blocks. Formed by the cupped palms of a silverback rapidly striking his body, the noise of chestbeats can travel more than a mile, but this one was close, rising out of the trees just across the clearing. It came again, a staccato challenge that hung in the air until someone from Katendegyere group pounded out a response.

"*Poc-thud-poc-thud*," the rhythm sounded strangely off-beat. *It's Kacupira*, I realized with affection. His broken wrist had healed long ago, but its sharp backwards angle prevented the left hand from striking palm-side down. The resulting syncopation identified him as surely as a fingerprint.

I saw the trackers smile with recognition, and Charles thumped his chest with one limp-wristed hand. We closed in slowly on Katendegyere group and the air grew thick with the pungent, armpit stench that told us the gorillas were nervous. Then another chestbeat echoed across the valley, crisp and resonant.

"That must be Makale or Mutesi," I whispered to Charles. "Coming to visit Katendegyere. This could get interesting."

As if reading from the pages of a gorilla behavior textbook, my two favorite bachelor apes had started to grow more bold. Where Makale and Mutesi used to run at the first sound from another silverback, they now began to seek them out, testing the limits of their new independence and power. Last week we found them shadowing Mubare group for three days,

never confronting Ruhondeza, but staying in the neighborhood and keeping him and his family in a constant, nervous retreat. Eventually, I knew
their posturing would lead to a serious encounter, where charges and displays of aggression could attract females to their cause, forming the
nucleus of a new gorilla family.

We moved forward in a tight group, beating a path through the foliage.
The forested ridge-tops of Rukubira and Hakanyasi rose up on either side
of us, leaving a shallow, bowl-shaped valley choked with thickets and
copses of tall canopy trees. In a trampled clearing we caught up suddenly
with Mugurusi, Nyabutono and Kasigazi. The 'old man' turned and
regarded us with a cool stare, his long face inscrutable, but calm. Light
rain clung to the black hairs of his coat in tiny droplets, reflecting the
cloud-light in a fine patina of silvery dew. He crossed the clearing in
purposeful strides, and headed up the far slope with the others close
behind. Kasigazi looked large and out of place riding on his mother's back.
At more than three years of age, he was almost fully weaned and
independent, but Nyabutono still kept him close when the group was
moving quickly.

They soon drew ahead of us on the hillside, and we decided to head
back towards the noise of Kacupira's chestbeats, hoping to witness a rare
inter-group encounter. Approaching from above, we found him and
Katome standing guard in a sloping patch of grass and low shrubs. They
looked out across the bowl of the valley where the bachelor males sounded
off again, closer now, just out of sight behind the trees. I watched
Kacupira purse his lips and begin a low croon. The sound rose into a series
of soft, chimp-like hoots as he reared upright, threw a handful of
vegetation into the air, and whacked his chest with four percussive slaps.
He then lurched sideways towards Katome and hammered at the ground
with both fists. The smaller ape side-stepped Kacupira's rush and edged up
the slope towards us. He paused to feed for a moment, channelling his
obvious anxiety into a non-threatening behavior. The technique can dispel
tension in minor conflicts, but Katome hardly even chewed the leaves

before continuing past us into the thick brush, following the same uphill route as Mugurusi and the others.

Kacupira ignored his companion's departure and crooned the beginning of another chestbeat sequence. With nine individual steps, from the vocalization, to tearing vegetation, running sideways and slapping the ground, gorilla chestbeats represent one of the most complex visual displays in the animal kingdom. All gorillas imitate aspects of the ritual, and we often laughed at the sight of juveniles like Bob or Kasigazi standing upright and thumping their bellies. But only adult males complete the entire display, using it in situations of extreme anxiety or outright aggression. As a precursor or companion to charging, the chestbeat can help competing males judge each other's strength and avoid direct physical conflict. The audible portion alone alerts any other groups or solitary males to a silverback's presence in the area.

With Makale and Mutesi approaching from below, Kacupira used the chestbeat both to answer their challenge, and as a physical outlet for his growing stress. Lunging at Katome had been incidental; his aggression was focused down the hill. I tried to remind myself of this as he retreated quickly towards us across the clearing, hooting, ripping at the shrubs, and preparing for another chestbeat. "*POC-THUD-POC!*" Even with a lame hand, the noise was powerful five feet away, and I could feel his heavy steps reverberating deep in the ground. He lunged closer, but still stared back over his shoulder, looking downhill for his real adversaries. We backed slowly into a thicket as he reared up again, lips parted over teeth like yellowed ivory. Standing nearly six feet tall, Kacupira probably weighed four hundred pounds and looked invincibly strong, but his brown eyes revealed something like panic as the object of his attention finally emerged from the trees.

Makale strutted into the clearing, stiff-legged, like belligerence incarnate. He looked larger than the last time I'd seen him, with the first grey hairs coloring the saddle of his back. Through binoculars I recognized his noseprint and steely glare as he scanned the clearing and advanced slowly into the open. We took advantage of the distraction to scramble further from Kacupira, but we needn't have bothered. He took one look at Makale and abandoned his sentry post without a single charge or chestbeat, disappearing up the slope behind the rest of the Katendegyere apes. Makale kept up his leisurely, steady pursuit, and we could hear Mutesi bringing up the rear, still out of sight in the trees.

Heading back to camp in a heavy downpour, we listened for chestbeats or vocalizations from the valley, but the gorillas must have sought shelter from the rain, putting aside their aggression, at least temporarily. Katendegyere group would probably continue their retreat, since Mugurusi had nothing to gain from a confrontation. Losing Nyabutono to Makale or Mutesi would leave him without a mate, and the rest of his group might well disperse.

I returned the next morning and followed the trackers to a wide field of flattened vegetation, site of an obvious confrontation. Trying to picture the encounter, we scanned the ground for signs of blood or fistfuls of hair, but found nothing. Apparently, charges and displays had been enough to settle the issue without coming to blows. But the interaction sent Katendegyere group running. Their trail led away in a single-file sprint through the forest as they fled from their persistent, uninvited guests. We counted night nests and knew that the group was still intact, but didn't catch up with them for two full days.

Chapter XVII

Some Grains Will Perish

—————————▼—————————

*"Some men have no children, and some grains perish
without fruit; then all are ended."*

—Commoro, Sudanese chief, *1862*

When the rangers found it, the hunting dog staggered away through the undergrowth, blind with pain. Its right rear leg was broken and its jaw hung useless, half-torn from the bottom of its ruined face. They tied the dog firmly with a bark-rope tether and led it back to headquarters. Later, people in the village identified its owner, giving the park its first lead in the investigation. They found another dog dead at the scene, killed in the fray and abandoned near the rotting corpses of its quarry: a black-back, an adult female, and two juvenile mountain gorillas from Bwindi's Kyaguliro family.

The rangers made careful note of their location, returning the next day with Liz and a team of investigators to identify the remains and conduct autopsies. Everyone brought cigarettes for the work, smoking furiously in a vain attempt to cover the stench of putrefaction. But coping with the smell was almost welcome, a visceral distraction from the more disturbing defeat and sadness of losing precious gorilla lives.

The poachers had attacked en masse, harrying the apes with dogs while they edged close enough to slash and jab with their long spears. No trophies were taken, but census reports of the Kyaguliro survivors showed an infant missing from the group. Its mother, the slain female, lay slightly apart from the other bodies and she probably died last, still struggling to protect the baby her assailants had come to steal. Gorillas defend their young to the death, and poachers always kill several other group members when they target an infant for capture. Although most zoos now refrain from purchasing wild-caught apes, gorillas still bring a high premium in the illegal pet trade. Collectors smuggle dozens of lowland gorillas out of West Africa every year, but this attack marked the first time in more than a decade that poachers had struck at a habituated family from the Bwindi or Virunga populations. If it survived, the baby abducted from Kyaguliro group would become the only captive mountain gorilla in the world, a tragic centerpiece for someone's private menagerie.

News of the attack reached Buhoma on a sunny morning at the end of the dry season, the same day we expected an official visit from Uganda's new Minister of Tourism and Wildlife. Our warden, Ignatius Achoka, was beside himself.

"You have heard of the poaching?" he asked me. "They have killed gorillas at Ruhija. Now how am I supposed to tell the Minister?"

To make matters worse, circumstantial evidence suggested that the poachers had help from the inside. Kyaguliro group was being habituated for a behavioral study by employees of the Institute for Tropical Forest Conservation (ITFC), a research station at the park's Ruhija ranger post. They visited the gorillas every day, just as we tracked Mubare and Katendegyere groups. Their presence should have made Kyaguliro one of the best-protected groups in the park. More than twenty unhabituated gorilla families ranged through the forest in relative obscurity, so why would poachers risk immediate detection by attacking such a well-known group?

Speculation and rumors implicated everyone from local rangers to the chairman of the Uganda National Parks board of trustees, but the most damning evidence pointed towards ITFC and its controversial director. Strangely, he had dismissed the Kyaguliro group's regular habituators only a week before the killings, replacing them with a team of strangers that he hand-picked himself. And when rangers examined the dead gorillas, they found them partially decomposed, as if ITFC had waited for several days before reporting their deaths to the park.

"He is very powerful, and he has sworn to take us all with him when he goes," a Bwindi warden told me. Under charges of corruption, the ITFC leader was in the midst of a bitter evaluation by The World Wildlife Fund, the Institute's major sponsor. "Maybe he decided to take some few gorillas too. I don't know."

Although respected as a scientist, the director had mismanaged ITFC to the point of bankruptcy, bringing its research activities to a virtual standstill. Students accused him of stealing their stipends, and spending the Institute's money as his own. At workshops and conferences, I remembered him as a knowledgeable, friendly participant, but those who knew better warned me of another side to his character.

"You must watch yourself with that man," an official at National Parks headquarters told me once. "On the outside he appears so clean–a born again Christian; he doesn't drink; takes no meat. But you only scratch the surface and he is black, black as coal."

Starting from the injured hunting dog, police and wardens identified five local suspects in the case. They raided their village homes and found large parties in progress, evidence of a suspicious influx of money into the families. But the poachers themselves had already fled into Zaire, leaving a host of unanswered questions in their wake. Who had hired them? Where was the missing infant? Who warned them of the investigation? Special agents arrived from Kampala and the inquiries dragged on for weeks, but no one at the park or ITFC was ever charged.

Whoever organized the incident had the connections to pinpoint a known gorilla group deep in the forest, locate and hire five local hunters, and smuggle an infant gorilla out of the country. Unfortunately, people with that kind of influence probably also had contacts in the higher levels of government. The ITFC director, for example, knew the chairman of the National Parks board, who was a personal friend of President Musevini. Through the transitive property of power, an important political force in Uganda, he was virtually untouchable.

Some people believed that politics played an even larger role, that the gorillas were killed in a ploy to embarrass and discredit the new Minister of Tourism and Wildlife. As an ex-general, and the former leader of a rebel group that once rivaled Musevini's National Resistance Army, the Minister had plenty of well-connected enemies in government. To lose five valuable mountain gorillas insulted him in his new role, a suspicious coincidence on the day of his first trip to the park.

Theories grew more speculative as it became apparent that the case would never be solved. Philippine authorities confiscated an infant gorilla at the airport in Manila, but it turned out to be a Western lowland. Rangers apprehended a poacher from the right village, but he had a credible alibi and knew nothing about the Kyaguliro incident. After several months, only a junior warden remained on the case

"They were just out hunting bushpigs," Phenny concluded one day, throwing up his hands in disgust. "We shouldn't let anyone into the forest. No tourism, no multiple use, no beekeepers, nothing. This is what happens."

Regardless of who was to blame, losing the Kyaguliro apes shocked everyone involved in the world of mountain gorillas. For years, poaching in Bwindi and the Virungas had focussed on small game—antelope and bushpigs that hunters hoped to add to their cookpots. Direct hunting of mountain gorillas had all but disappeared since the mid-1980s,[*] allowing park managers to focus on more far more far-reaching conservation strategies, combining protection with tourism, conservation education, revenue sharing, and other progressive ideas. In Bwindi, we planned for sustainability, thinking in terms of economics and genetic viability over the long term. The threat of gorillas dying at spearpoint had seemed remote, a problem left securely in the past.

The Kyaguliro killings brought that menace back to the foreground, a brutal and unsettling reminder that despite our best efforts, mountain gorillas still faced the risk of immediate extinction. I no longer took it for granted that Mubare and Katendegyere groups would be waiting safely in the forest each morning, and found myself questioning some fundamental aspects of the park's management plan. Our program emphasized the benefits of conservation over time. Through education, tourism, and local involvement, we hoped to show people that saving Bwindi forest and the gorillas would improve their quality of life. But how well had these concepts taken hold if villagers and those behind them were still willing to hunt and kill the apes?

By definition, sustainable strategies depend on communicating the idea of future rewards. Local communities and governments give up the short-term gains of logging and hunting in return for a continued yield of

---

[*]   Wire snares set for small game still pose a threat to gorillas. While the apes can easily break free, the snares cinch deep into the flesh of hands and feet, causing infection, maiming, and occasionally death. The Volcano Veterinary Center, based in Rwanda, also works in Zaire and Uganda, removing snares and treating any other human-induced wounds of gorillas in the wild. Liz Macfie led that project for three years before moving to Bwindi with IGCP.

Also, while direct poaching is rare in mountain gorillas, their lowland cousins continue to suffer terribly. In West Africa, locals regard both gorilla and chimp meat as a delicacy, and commercial hunting fuels a growing bush meat trade as rural populations and logging companies encroach further into the rainforest. Gorilla parts are also used in a variety of local fetish medicines, i.e. to imbue a person with the legendary strength and virility of a silverback.

tourism revenues and environmental benefits. A potential problem, I realized, lay deeply rooted in the African sense of time, a framework that emphasizes the current moment over what is to come. Rural cultures had evolved in a kind of stasis, with few dramatic variations in weather, agriculture, or the pattern of daily life. Without any anticipation of change, the future became almost irrelevant, a simple extension of present conditions. Its low priority is revealed in language, where several different verb tenses describe the past, but only one exists for future times, a vague 'tomorrow' indicating nothing beyond the next several days.

People live continuously in the present, a state of mind I came to understand while working with the trail crew. During the wet season, moss-heavy branches and whole trees fell constantly across our hiking trails. I struggled for months to teach the idea of maintenance, but still the men would step over the logs rather than removing them from the path. To me, taking a moment to clear windfalls saved the effort of climbing over them countless times in the future. To the trail crew, however, the idea was a ludicrous waste of energy. If you consider walking the path as an event isolated in the present, then it's far more logical to step over the log, rather then spend twenty minutes chopping through it and pushing it out of the way. In the absence of future, my drive to keep the trails clear must have seemed slightly insane.

Using the same model, abstract rewards like revenue sharing or water-shed protection may never outweigh the immediate benefits of hunting, cutting timber, or selling an infant gorilla. In fact, they might not even make sense. The trail crew eventually became highly skilled in mainte-nance, but I was never sure whether or not they agreed with the concept, or if they did it just to humor me. Similarly, the strategies of sustainable management may someday produce great results, regardless of the motive, and that potential makes any amount of effort worthwhile. But seeing the rebirth of active poaching increased my capacity for doubt.

Five months after the Kyaguliro incident, poachers shot and killed three more gorillas in Zaire's Virunga National Park, including the lead silverback of their most popular tourist group. The killings were probably unrelated to those in Uganda, but brought our losses for the year to eight individuals, nearly one and a half percent of the entire mountain gorilla population. Where we had once shunned firearms on tracking duty, we now posted armed ranger patrols to follow Mubare, Katendegyere, and the remaining Kyaguliro gorillas throughout the daylight hours.

((((

I knew the route to the herbalist well, bumping along the narrow path and braking automatically to swerve past the goats, cows, and pot holes. Phenny and Hope rode with me, our conversation interrupted by long stretches of concerned silence. Amanda lay still in Hope's lap, hazy-eyed and despondent, her tiny face pinched with sickness. In the wake of the gorilla poaching, it seemed the season for sadness had returned.

"We didn't know they cursed the baby too," Phenny would tell me later, wooden with grief. "We didn't know."

I pulled into the yard and Hope stepped carefully down from the vehicle, bundling the child against her shoulder. Amanda's head looked too large and heavy for her thin little body, lolling loosely backwards as Hope hurried inside.

Phenny and I drove back to the market for eggs and oranges. "Hope always wants them," he said, piling the tart green fruit into a basket. He bought enough food for a long stay, but the visit turned out to be brief; Amanda died while we were shopping.

At the funeral, Karen and I mingled with hundreds of guests. Everyone in the village seemed to be there, milling through the compound, talking quietly, and singing atonal hymns by the grave site. As patriarch, Phenny's father Tibesigwa presided over the service and we paid our respects to him before stopping to sit awhile with Hope. Phenny drifted through the crowd alone, hollow-eyed, shaking hands and greeting people with a mechanical smile.

The atmosphere was solemn, but social, and I saw most of the guides and trackers. William Betunga joined me for a walk through the banana *shamba* to the burial plot, a small family cemetery tucked into the Gongo's hillside farm. Amanda's grave was a low mound of freshly-turned earth, strewn with colored flowers. We said nothing, but I knew William mourned for the child. He and Phenny were neighbors and inseparable friends. Without a family of his own, William often took meals with the Gongos, and treated Amanda as a favorite niece or sister. I remembered him holding her naked on his lap, bouncing to the music from his Walkman radio.

"Maybe she will shit," he'd said happily, grinning a hopeful smile. "If a baby shits on you, it means you will have many children of you own!"

Nearing thirty and still single, William's lack of a wife and children was anomalous in the village. As a rule, Bakiga men married in their late teens, or as soon as they could afford to. They looked upon raising a family as something fundamental, a quantitative measure for the success of their adult lives. In the largely static economy of the village, children represented a certain wealth: help on the farm and security in old age, as well as the economic potential of a son or daughter finding good work in the city. But the motivation for large families ran far deeper, to the very roots of local culture.

"People pray to their ancestors," Enos Komunda once told me, explaining how traditional religions revolved around consultation with the dead. The power of individual spirits related directly to the size of their family clan, and the length of time they'd been deceased. "For a small thing, you might pray to a grandparent. But for larger problems, people had to look further back." Ancient spirits held sway over a vast genealogy of the dead and living, the same way village elders supervised their clans and families. This influence increased with age until those beyond memory finally merged with God, the ancestor to all.

Culturally, children represented the link between future and past, a legacy of primary importance. In this passage from his journey in 1862, explorer Samuel Baker describes a conversation with Commoro, chief of the Lakoota tribe in southern Sudan. Baker tries to explain the Christian concept of afterlife, and Commoro replies afterward with an elegant summary of the African perspective.

> *"Some corn had been taken out of a sack for the horses, and a few grains lay scattered on the ground; I tried the beautiful metaphor of St. Paul as an example of a future state. Making a small hole with my finger in the ground, I placed a grain within it: 'That,' I said, 'represents you when you die.' Covering it with earth, I continued: 'that grain will decay, but from it will rise the plant that will produce a reappearance of the original form.'*
>
> *Commoro: 'Exactly so; that I understand. But the original grain does not rise again; it rots like the dead man and is ended; the fruit produced is not the same grain that we buried, but the production of that grain. So it is with man—I die, and decay and am ended; but my children grow up like the fruit of the grain. Some men have no children, and some grains perish without fruit; then all are ended.'"*

Without children, then, people feel an acute sense of impermanence, and no connection to the spiritual continuity of their forefathers. To die in such a state is to be unremembered, with no link to the history of one's family and clan.

In an era of improving public health, this cultural imperative leads naturally to population growth, and the number of people in southwestern Uganda is expected to double over the next fifteen years. More subtly, however, it can also contribute to the spread of disease. A person who tests HIV-positive might not choose to abstain from sex. In fact, they might try even harder to reproduce, desperate to leave their progeny behind as a connection to the world of the living.

William and I stood together in silence beside Amanda's tiny grave. I looked for a flower to add to those scattered over the mound, but the bougainvillea was already stripped bare, and the bananas shaded out anything wild.

That evening, I met the ranger patrol returning from Katendegyere group. They told me they'd left the gorillas feeding in the Muzabajiro Valley, just over the hill from my house. On impulse, I set off up the trail, crossing into the forest through the notch above Dominico's place. The trees dripped from an afternoon rain squall and I could hear the faint drumming of the creek, still swollen with runoff. My path led along the side of the narrow valley, a ledge carved into the hillside and lined with wild Impatiens. I walked slowly, listening to the day's last birdcalls: tinkerbirds, a tchagra, and two cinnamon-chested bee eaters, rasping their three-note whistle and darting over the treetops like jade swallows.

I heard familiar grunts as I approached the gorillas, and coughed lightly to make my presence known. They were bedding down just above the trail, in a steep, shrubby clearing laced with white blossoms. I climbed a few meters towards them and sat quietly in a patch of wet grass. Thick bracken blocked the apes from view, but I heard the snap of branches as they settled in for the night, weaving their nests around them. The three close together would be Mugurusi, Nyabutono and Kasigazi, with Katome somewhere nearby. Higher up the slope I heard the others—Karema, probably, and Kacupira, who lived more and more on the periphery these days, as if constantly keeping watch for Makale and Mutesi.

Reluctant to leave them, I lay back against the hillside and felt the cool earth dampen my shirt. A soft belch from up the hill, followed by rustling leaves as someone turned in their nest. And then silence, a damp hush that swelled with chirring cicadas as daylight faded quickly from the sky.

CHAPTER XVIII

THE SOUTH SIDE

▼

*"Were the gorillas on the three peaks protected I am certain that
in a very short time they would become so accustomed to man
they could be studied in a way that would rapidly produce the
most interesting and important results."*

—Carl Akely, *1920*

"Mr. Tour," Kabahoze Fred called, tapping my shoulder and holding out a chunk of black plastic. "From your boot."

"Thanks, Fred," I said wearily and tucked the piece into my pocket with the others. Looking down, I saw a new hole near my left big toe. The soles and both insteps were already split, and the front half of the boots flopped open with every step, exposing my wet socks to the mud and thorns.

This is over-doing it, Jungle man, I thought, and kept moving up the hill. I should have replaced my leaky rainboots long before setting out on a five day tracking journey. But that wasn't my only mistake. At camp the first night, Fred had looked up worriedly from the cookpot.

"Did you bring salt?" he asked, as the others gathered round to eat.

"No," I said in growing horror. "You mean we don't have any salt?" When your only food for five days is boiled beans and corn flour, forgetting the salt is a major culinary setback. The other trackers hung their heads.

"But...a man cannot live without salt," lamented Benjamin, turning to Gaston for support. As veteran rangers, they'd both spent countless nights in the forest on anti-poaching patrols. But never without salt.

Gaston only shrugged, and carefully pulled a wild red pepper from his pocket, setting it close to the campfire to roast. Laconic and permanently smoky-eyed, he had a reputation as the best woodsman in the park. Watching enviously as he stirred hot peppers into his bowl of beans, I was ready to agree.

We spent the night near the rushing Kashasha River, at the foot of Kasatoro, an 8,000 foot ridge near the park's southern boundary. Benjamin, Fred, and Gaston worked as part of our new habituation team, assigned to locate a third gorilla family suitable for Bwindi's tourism program. We had postponed this phase of the project after the poaching of Kyaguliro group, reevaluating the risks and benefits of habituation. Historically, tourism and research had always helped reduce poaching, logging, and other illegal activities. The daily presence of rangers and trackers drove most hunters away, making tourism areas among the safest in the park. But when that visible protection wasn't enough, determined poachers would find habituated apes an easy target, as they had with Kyaguliro and the recent slayings in Zaire.

In the end, the decision came down to fundamentals. Tourism provided Bwindi with revenue, and revenue made the park economically important, both locally and nationally. Losing that incentive posed an even greater threat to gorillas: destruction of the forest itself. We stood by our original concept for the tourism program, that three habituated gorilla families could provide enough income and interest to help sustain the park in perpetuity.

The challenge now lay in finding an accessible group. At dawn we drank cups of hot corn porridge and started up Kasatoro's endless slopes, following a day-old trail of the Nkuringo family, fifteen gorillas named for a dome-shaped hillock halfway up the ridge. The park's head warden, Ignatius Achoka, had mapped their range as part of his Masters thesis.

"The place is steep," he told me, making a vertical line with his hand. "But they should be the closest."

Climbing Kasatoro, with my boots disintegrating from my feet, I wondered if these gorillas were close enough. Our base camp lay a full day's hike from Buhoma, and now we'd added seven straight hours of uphill tracking. My legs felt like bent coat hangers, and I'd been doing this kind of thing for two years. Tourists would need a chairlift.

"The trail is still old," Benjamin concluded, rising up from a quick examination. "Do we continue?"

"Sure," I sighed, and drank from my water bottle. "As long as we can get back by dark."

"Down is fast," he assured me, and scrambled up the path out of sight.

At last we reached the night nests, a loose group of matted leaf and twig structures tucked into a narrow draw near the ridgetop. We paused to count them, making sure that at least we'd spent our day following the right group. Gaston, Fred and I split up, marking each nest with an upright branch while Benjamin came along behind to count and measure the dung. He pulled up our sticks as he went, making sure that no nests were counted twice.

"Fifteen, with four males," he concluded. "This is Nkuringo."

We pressed on, scrambling up an old creekbed as the sun dipped orange towards the west. Scattered clouds threw slow-moving shadows across the ridge and we hiked through patches of dimness. At this altitude, the forest gave way to frequent clearings, and old rock slides choked with ferns and low shrubs. The noise of our *pangas* startled up several duiker, spaniel-sized forest antelope with tiny straight horns and coats the color of fall leaves. They dashed away in panicky leaps, their nasal alarm cries like shrill whistles or people blowing party favors.

Moments before I called a halt for the day, Gaston suddenly came to life, scanning the ground with languid interest and moving up for a whispered conference with the others.

"The gorillas are very near," Benjamin translated. "If we hurry we can still reach them."

I nodded and we picked up the pace, racing against the threat of darkness. Fred took the lead, and I watched him choose our route under Benjamin's critical tutelage. As the most inexperienced of the trackers, he was still learning about the art of following gorillas.

Without warning, a silverback charged us from up the slope. I glimpsed a flash of its screaming blackness, and then Fred's terrified face as he streaked by me, sprinting away down the hill. *Never run*, I reminded myself, crouching automatically as the gorilla bore down through the foliage. While charging is primarily a display and a bluff, the apes will follow through if their opponent flees, knocking him to the ground from behind. Not surprisingly, most gorilla-related injuries involve a nasty bite to the backside.

Gaston yanked Fred down as he sped past, and the silverback retreated immediately, disappearing uphill in the dense greenery. We settled into the mud and ferns, waiting for another charge. Suddenly, a swarm of biting flies descended upon us, wave after wave, like nothing I'd ever seen. They covered every inch of bare skin and worked themselves up pant legs, through boot holes, and down our shirt collars. The stinging itch was unbearable, but any sudden movements would only antagonize the nervous gorillas we'd come to habituate. I looked at the others and raised my hands in the universal sign for *'you've got to be kidding me!'*

Benjamin shrugged back and Gaston gave me a resigned, half-smile as he slapped at the bugs. Fred was too embarrassed by his panicked flight to meet my gaze, and looked sheepishly at the ground while we waited out the attack, wiping handfuls of flies from our faces and arms like tiny, buzzing gravel.

I heard faint sounds of movement from up the hill and we edged slowly closer, but the silverback had already departed, following his family towards the top of the ridge. For the initial stage of habituation, this was a classic response: the rear guard charges while the rest of the group moves quickly and silently away. We tracked ten hours for a ten-second interaction, but even a single charge was better than no contact. Nkuringo group lacked any experience with people, and only long months of daily repetition would convince them that our presence was benign. Eventually, they might grow calm and curious like Mubare group, or learn, like Katendegyere, to anticipate our arrival and actually use it to their advantage.

On a recent visit, we had found the Katendegyere apes waiting for us at the forest edge, just across the border in Zaire. Kacupira stood at the rear of the group, poised for travel. He looked back at us over his shoulder, barked at the others, and they all took off, racing across a ravine and up into a neighboring banana *shamba*. Mugurusi led the raid, a line of six sable shadows moving with hurried purpose through the short grass and banana stems. They knew that for the full hour of our visit, they could loot the fields without fear of retribution. With their park ranger friends standing by, no farmers would dare throw rocks or try to chase them away.

We crossed into the *shamba* and saw Karema, madly stuffing handfuls of banana fiber into his mouth and gulping them down. He didn't even glance up from his meal as we approached within twenty feet and stood watching. But when two kids from the farm shouted to one another far up the slope behind us, he dropped his banana stem and sat up straight, peering warily over our heads for any sign of the strangers. Seeing nothing, he redoubled his efforts on the feast, chewing frantically until his chin, chest and forearms were drenched with sticky juice.

I told that story to Benjamin as we made our way down Kasatoro in the late evening light. He said that Nkuringo group also raided the banana *shambas*, but never let people come close enough to watch.

"But we will habituate these gorillas," he assured me with quiet confidence. "It will only take time."

I left the Nkuringo trackers hard at work and returned to other tasks in Buhoma. "Come to my *obugyeni*," I said in parting, inviting them to my farewell celebration the following month. "We will have *matoke, tonto,* and even two goats!"

Benjamin accepted for all of them, then gave me an important tip for the menu. "Salt," he said solemnly. "This time, don't forget the salt."

CHAPTER XIX

THE IMPENETRABLE FOREST

▼

*"...Bwindi has become the main place in East Africa for
seeing the mountain gorillas. A major conservation effort has
been going on here for a number of years to protect the gorillas'
habitat. As a result, encroachment on the montane forest by
cultivators has been stopped, poaching has ceased and the gorilla
families have been gradually habituated to human contact..."*
—*Lonely Planet Guide to East Africa*, 1997

Ruhondeza rolled onto his back and passed gas, a long, raspy sound,
like the coughing start of an outboard motor. I saw the tourists smile and
laugh as they realized what they were hearing.

"This is common?" whispered a tall man next to me. He spoke in a
clipped, German accent and sounded offended, as if surprised that such a
noble symbol of the wilderness could be capable of flatulence.

I told him that it happened all the time, one of the less glamorous
aspects of gorilla behavior that gets edited out of National Geographic
specials. With a fibrous fruit and vegetable diet, and their own collection
of stomach parasites, mountain gorillas surpassed even Peace Corps
volunteers as the gassiest primates on the African continent. As any

tracking veteran can tell you, a good portion of gorilla-viewing involves sitting around in the rain, listening to the apes break wind.

We'd found Mubare group feeding on broad-leaved *Brilliantasia* shrubs less than half a mile from my house. The plants flowered only once every four years, brightening Bwindi's dim understory with a brief flurry of white, like tiny lanterns in a hanging garden. The gorillas relished these rare blossoms, but perhaps more for their taste than for aesthetic reasons. I watched Ruhondeza pull down a branch and chew lazily at the dangling blooms, like a Roman god nibbling on vine-ripe grapes.

Lying at the opposite end of the habituation spectrum from the Kyaguliro gorillas, Mubare group had become nearly indifferent to people and sometimes lingered for long periods near the busy footpath that connected Buhoma to villages on the south side of the forest. I'd spent several market day afternoons escorting groups of fearful locals past the apes. The situation provided an excellent opportunity for conservation education.

"*Twareeba enjojo*," whispered one wide-eyed youth: 'We have seen the elephants.'

I wanted the community to learn more about gorillas, but worried that curious apes could expose themselves to unnecessary harassment and disease risks. Properly done, the habituation process shouldn't simply eliminate fear, but replace it with respect. We aimed for peaceful coexistence, with both humans and gorillas learning to maintain a safe distance. But like Makale in Katendegyere group, the Mubare gorillas, particularly the juveniles, always found new ways to challenge our teaching.

I crouched next to Tibamanya, the lead tracker, and scanned the group with binoculars. Mamakawere reclined near Ruhondeza with her baby cradled against her chest. At five months of age, the infant's skin had darkened completely, transforming the fragile newborn into a tiny, rounded replica of its massive parents. Motherhood seemed to give Mamakawere increased status in the group, and we rarely found her far from Ruhondeza's side.

Further afield, three young apes played king-of-the-mountain on a low stump, grappling, biting, and shoving each other roughly to the ground. Bits of mud and dry leaves hung from their long black hair, and their soft play hoots sounded like cartoon monkeys as they struggled to climb higher. The largest finally prevailed, pushing away its siblings with a heave, and pirouetting awkwardly on top of the mossy stump. It sat down, licked rainwater from its forearm, and peered around with youthful distraction. When that brown-eyed gaze fell on us, the ape sat up with sudden interest, then dropped to the ground, staggering in our direction on two legs like a drunken toddler.

For the young gorillas in Mubare group, trackers and tourists blended into their daily landscape like the trees and shady clearings of the rainforest. They accepted human visitors easily, a predictable occurrence of quiet, camera-toting observers they'd seen on every day of their lives. As the apes matured, their natural curiosity made people an irresistible subject for investigation, sometimes at close range.

As the youngster came towards us, we gathered the tourists together, preparing to retreat and maintain our five-meter distance. This tactic failed however, when a gorilla was persistent, or drove us into thick vegetation. Instead, we'd begun shaking leaves and twigs at the juveniles whenever they crossed the five-meter line. This motion and noise irritated them visibly, and usually drove them back, but we had to be careful. If annoyance turned into fear, they might cry out and bring down the wrath of the rest of the group.

I plucked a handful of ferns and stood firm with Tibamanya. The young ape advanced on all fours now, pausing after every few steps to peek cautiously through the undergrowth. From its size and independence, I judged its age at nearly three years. Within twelve months it would separate completely from its mother and begin building its own night nest, marking the definitive end point of its gorilla childhood. The female would come into estrus again soon after, continuing the gorillas' constant reproductive cycle: breeding, an eight and a half month pregnancy, and three to four years of intensive child rearing. While infant mortality ranged as high as forty percent, a healthy female might still raise four to six children in her forty-year lifetime.

Young gorillas receive more parental care than almost any other primate. Born light-skinned and highly dependent, infants rarely leave their mother's side during the first six months. They ride ventrally, clinging to her chest and nursing at regular intervals while they gain in size and dexterity. Gradually, they learn to perch on her back for long journeys, but don't start travelling on their own or interacting with other juveniles until well into their second year. Silverbacks often take an active role in child-rearing, and have been known to 'adopt' juveniles whose mothers have died. In most primate species, mating with multiple males makes parentage unclear, and child-rearing is left entirely up to the females. But as the sole breeder in his group, a lead silverback like Ruhondeza can treat all the offspring as his own, and take a greater interest in their care. We often saw him playing and wrestling gently with his younger children, and once, when a two-year-old fell from a high branch, Ruhondeza was there, stretching his arm up casually to make an easy, one-handed catch.

Tibamanya and I began rattling ferns and leaves at the approaching juvenile, and it stopped suddenly, blinking and wrinkling its nose, as if the noisy vegetation were some kind of bright, foul-smelling flashlight. We paused, and it twitched its head from side to side, peering up at us with a suspicious, narrow-eyed look before taking another step closer. I shook the ferns again and it spun away, then came at us from a different angle. We continued this game for several minutes until the ape grew bored, and turned back to rejoin the group. As it walked slowly away, I pulled out my notebook and tried to sketch a noseprint. With so many look-alike females and youngsters in the group, we'd named only a few of the Mubare gorillas, but today's visitor seemed distinctive. It stood noticeably larger than the other juveniles, with a heavily lined face and a precocious capacity for mischief.

"Does this one have a name?" I asked Tibamanya quietly.

He held a whispered conference with the other trackers, and turned back to me, nodding. "Bob," he whispered.

"Bob?" I was taken aback. Ruhondeza, Kashundwe, Muchuchu,…and Bob? Where did they get this name, I asked. From a tourist? No. Did it mean something in Rukiga? No. Then how did they choose it?

Tibamanya shrugged, blinking his perpetually sleepy eyes. "He is just Bob."

Strangely enough, the name seemed to fit. I watched the young ape pick up speed and launch himself onto his father's stomach, eliciting a low growl from the resting silverback. Ten minutes later, the whole group moved off, with Bob bringing up the rear. He glanced back over his shoulder with a puzzled expression, as if wondering why we all didn't hurry to catch them up.

((((

Over the past two years, Mubare and Katendegyere groups had received more than five thousand visitors, from local people to foreign tourists and visiting dignitaries. As Bwindi's reputation grew, it attracted attention from far beyond the community of adventure travellers. The World Bank identified Bwindi and nearby Mgahinga parks as beneficiaries of a perpetual trust fund to support protection, research and sustainable development. Sheraton Hotels donated construction funds and sponsored a design contest for a new Bwindi visitor center. CARE International hired a park advisor and stepped-up its sustainable development projects in the area, while Liz was hard at work on a major grant proposal that would extend IGCP's support for three years, bringing new vehicles, equipment and additional staff for the park.

In many ways, the Impenetrable Forest had become a test case for modern theories of conservation. With gorillas as the focal point, Bwindi offered the opportunity to expand beyond protectionism, to see what ecotourism, revenue sharing, multiple use, and sustainable development could contribute to preserving biodiversity. Environmental groups and scientists around the world would watch Bwindi's successes and failures, taking the lessons home to their own conservation efforts.

All of this took place against the backdrop of Uganda's continued political stability and newfound position as a power broker in the region. While people criticized Musevini for his resistance to multi-party politics, all agreed that his government had brought about a dramatic social and economic recovery in Uganda. Aid poured in from developed countries, rebuilding infrastructure and attracting foreign investment. When Musevini visited Europe and the United States, he gained the reputation as an eloquent statesman among the new generation of African leaders. Unrest still marred the northern part of the country, and rebel groups staged occasional raids from neighboring Zaire, but the path ahead looked hopeful. And ultimately, a secure, stable nation was the best hope for the forest and the gorillas as well.

In Buhoma, I watched these developments with great encouragement, but also with a certain nostalgia for the days when hearing a vehicle in the village was unusual, and when you didn't have to wait for a table at Hope and Phenny's Canteen. Gorilla tourism had brought jobs and relative prosperity to the community, but not without certain costs: inflation at local markets, children begging from foreigners, insensitive visitors offending local customs, and even the arrival of prostitutes to serve the tour company drivers and camp workers. Still, it was an exciting time, with a new focus on the future.

"I am investing in wood," one of the guides told me. "If I buy timbers now they are cheap, but I'll sell them later for profit when everyone is building new houses!"

On a rare day without tourists near the end of my tenure, the trackers and I spent a quiet hour with Katendegyere group, sitting near them in the rain while they slept and fed. Only Kasigazi had shown any sign of activity, climbing a low tree to swing and dangle lazily from the branches.

As the group's lone juvenile, he'd learned to play by himself, wrestling with saplings, or making daredevil leaps from high trees. Before settling on the name Kasigazi, the trackers called him *Makyita*, a word meaning "courage and mischief." Today he'd hung one-armed from a stout limb, staring down at us with languid curiosity while he deftly picked his nose and ate it. *Now this is science*, I thought. Another kind of observation that doesn't cut it for the television specials.

Towards the end of the hour, Prunari touched my arm and pointed up. Twenty feet above us, a fine-banded woodpecker focussed his percussive attention on a rotten limb, then a liana, hopping nimbly from one to the next with no regard to gravity, as if he'd lost all sense of earth and sky in the serpentine weave of branches. His olive plumage blended perfectly with the leaves, a pattern of green broken only by the tiny crimson feathers streaking his crown. Through binoculars I watched him leap and flutter through the foliage, pausing to listen for the telltale gnaw of insects. He would cock his head and wait, silent as a still life, then rear back to hammer the bark away in sharp staccato bursts.

I thought back to my first months in Bwindi, when visiting Katendegyere group wasn't exactly peaceful enough for birdwatching.

Hiking home, the trackers skipped nimbly over rootwads and patches of rain-slick mud, half-running. It was market day, and they could sense the *tonto* flowing in downtown Buhoma. Overhead, sunlight crept through pillared gaps in the clouds and fell across our backs, searing hot after the rain. From Rukubira's hilltop clearing, the forest stretched away to the edge of vision in waves of textured verdure, heavy with steam and silence.

Unlike the teeming jungles of a Hollywood movie set, the heart of a true rainforest often seems deserted. As we descended from Rukubira, Bwindi closed around us like the dim hall of a medieval throne room, its wildlife hidden behind a living stillness that hung down from the canopy in veils of dripping green. The layers of foliage held back wind-hum and sun-bright air until any noise at all sounded thin and contrived: the monotone coos of

a tambourine dove, the bi-plane drone of Goliath beetles, or the subtle patter of fruit pith dropped by a troop of monkeys.

I stopped Prunari, and we advanced quietly to the base of two huge fig
trees, their wide trunks buttressed by smooth grey roots, like hip bones
sculpted from sand. High in the canopy, faint shadows coalesced into
monkey shapes, red-tails and blues, foraging together on the ripe green
fruit. They fed voraciously, ignoring their neighbors and the black-billed
turacos that hopped and flapped through the branches around them. Figs
and other fruiting trees are the market day *tonto* bars of the rainforest,
attracting a huge variety of birds, primates, and other species. By plotting
the location of specific trees, and recording the months in which they bear
fruit, researchers can predict the best places to find wildlife at any time of
year. We kept track of the trees near our forest walk trails, and this pair of
figs had been producing steadily for over two weeks.

When the monkeys finally noticed us, the whole troop erupted into
action, their sudden movements accompanied by panicky high-pitched
chirps. We mimicked their calls, cupping our palms against wet lips and
making sharp kissing noises. Several males paused to look down, bobbing
their heads as a gesture of threat. Then the turacos took flight—a flurry of
crimson wing-splashes that sent the nervous monkeys leaping away across
an aerial highway of branches and vines. Two redtails hurled themselves
haphazardly into dense thickets, a noisy display intended to distract us,
and alert any lagging group members to the threat of people below.
Historically, the Bakiga rarely hunted primates, but local Batwa pygmies
and many Zairian tribes relish the meat, and Bwindi's monkeys have
learned to be wary.

We walked on and saw the troop again, passing over a narrow clearing
in the forest. Each individual used the same route through the trees,
hurtling across a forty foot gap in the canopy like circus acrobats in
freefall. They jumped without pausing, arms stretched forward and tails
out for balance, their short-fringed coats glowing in the patch of sunlight.
The flashy russet fur of the redtails contrasted sharply with the blues, who

gleamed in shades of dull silver, like woodland spirits woven from mist. I watched them disappear into the greenery, sorry for disturbing their meal. But both species belonged to the genus *Cercopithecus*, a family of forest guenons equipped with large cheek pouches for storing food. At the first hint of danger, the monkeys had surely stuffed their mouths with fruit, a portable luncheon to enjoy in peace whenever the troop stopped running.

I let the trackers go on ahead and turned my walk home into a long, slow detour through familiar forest. A blue mother-of-pearl butterfly drifted above the path before me, furling and unfurling its iridescent wings like the folds of a magician's cloak. I recognized every bend in the trail as I moved towards Buhoma: a familiar tunnel through drooping Brilliantasia shrubs; the nesting tree for bar-tailed trogons; a day-roost snag for eagle owls. I passed the waterfall trail, where Phenny, John and I had hiked to the bridge on my birthday, dangling our legs over the stream in the black hush of midnight. Every branch in the path led to hills or tiny clearings where I'd camped, watched gorillas, or shared lunch with the trackers—obscure, but beautiful places with names like woodwind music: Hachogo, Kanyampundu, Rutojo, Mukeshwiga, Hakigugu, Musharara, Nyakagera or Hakatare.

The Peace Corps talks of community as the village world where volunteers interact with local people, form ties, and make a home for themselves. I realized now that my own community extended beyond the human element of life in Uganda. It included gorillas, clear streams, trees, birds, and all the myriad aspects of the forest itself, a dynamic, living landscape that I would miss as much as I missed my friends in Buhoma. I watched a rufous warbler flutter through the trailside bracken and felt a sudden wave of emotion, as if I were already surrounded by memory.

Chapter XX

Bye, Phenny, Bye

▼

My age-mates have donned
White ostrich feathers,
They are singing a war song,
I want to join them
In the wilderness
And chase Death away
From our village,
Drive him a thousand miles
Beyond the mountains
In the west,
Let him sink down
With the setting sun
And never rise again.

—Okot p'Bitek, Ugandan poet
from *Song of Prisoner, 1970*

"You have lost your friend," the man from Buhoma told me, and the chaos of the taxi park faded into a blur around us. "Hope is dead."

I nodded numbly, and heard myself ask about the burial.

"It should be tomorrow," he said. "They are still bringing the body from Kampala. You know she went for treatment to Mulago Hospital?"

I didn't know, but thanked him for telling me. He said he was sorry, and then disappeared back into the swirling crowd.

I felt disoriented as my surroundings came back into focus. A tiny boy next to me raised his eyebrows as I glanced around, holding out a box filled with gum and packets of sweet crackers. "Biscuits?" he asked, but I waved him away. Nearby, the driver of a bright green *matatu* honked his horn repeatedly, smiling and shouting like a drunkard. "*Muzungu*! Where are you going??"

Home, I thought with double meaning, and wearily began looking for transport. Ishaka's taxi park lay at a busy crossroads, but *matatus* rarely strayed from the main routes, and I knew I'd be lucky to reach Buhoma in time for the funeral. I'd been away for more than a week, working in Kampala and then stopping up-country to say goodbye to several Peace Corps friends. Our 'close of service' date was at hand, and we all had plane tickets back to the States. I'd planned a vacation in Zanzibar along the way, and for a shameful moment, I wished I was already there.

"I guess Uganda couldn't let me go without another round of tragedy," I later wrote in a letter home. Hope had battled poor health for more than a year, but somehow, her death still came as a shock. I think we'd all grown accustomed to the routine: small improvements and setbacks, new medicines, and trips to the herbalist.

"How are you, Hope?"

"Improving, Tour. I should be back to work any time."

Even sick, she kept herself heavily involved in community affairs, organizing meetings and managing accounts, while at the same time running her and Phenny's Canteen, the most successful local business in town. She treated her illness as a matter of course, and everyone took it for granted, until its implications almost ceased to seem threatening.

The struggle to reach Buhoma had never been so frustrating, and I didn't arrive until late the following evening, after a night in Kabale and a long series of buses, trucks and mini-vans. I missed the funeral service completely, but found the Gongo's farm still crowded with mourners as I made my way up the path.

"Tourrr!" Phenny called out when he saw me, drawing out my name like a wail. He ran down through the banana trees and shook my hand, his face unreadable.

"I got the news at Ishaka," I told him. "Someone found me there."

He didn't meet my gaze and we stood together for an awkward moment, but this time something greater than cultural boundaries kept us silent. In a span of months, Phenny had lost his only child and now his wife. Even in America, I wouldn't know what to say. "I'm so sorry, my friend," I finally managed, still gripping his hand.

He turned away up the hill. "You should greet my father," he said in a toneless voice, and led me through the lingering mass of guests. Tibesigwa Gongo sat with two of his brothers and a group of elders from the village, playing cards in the shade of a jack-fruit tree. I joined them, and Phenny moved off, talking quietly with several cousins from Butagota.

Mr. Gongo slapped down cards with concentration, playing *matatu*, a game I remembered from my days in Kajansi. "You know John DuBois was here when my wife died," he said without looking up. "Now you can stay the night for Hope's vigil. To remember."

Bright yellow weaver birds chattered and dove through the branches above us, and I thanked Gongo for the invitation. "I only wish you didn't have to have so many burials," I told him, and he smiled wanly, saying nothing.

Later, a brief cloudburst drove all of us inside. We stood shoulder to shoulder under the eaves, watching huge drops bounce like hail off the benches and hard-packed earth in the yard. Water hammered the metal roof like a timpani, and no one bothered trying to speak as the sky faded slowly to streaks of grey and twilight. I closed my eyes, breathing smoke from the cookhouse and the smell of damp bodies all around me.

People drifted away after the rain until only family and close neighbors remained, but we still numbered more than fifty, and Gongo had to cook a whole goat for the feast. We ate in shifts, the women gathering inside while the men sat around a huge bonfire, drinking hard and telling stories. Jug after jug of *tonto* appeared, and the atmosphere grew strangely festive as the night wore on—more like a stag party than a wake. I found myself at a long table playing cards with Phenny, William Betunga, Agaba Philman and others.

"Tour is the 'Asshole!'" they all shouted as I lost another hand. I laughed with them and shuffled the cards, silently cursing whatever tourist had taught them such an irritating card game.

All night long I kept an eye on Phenny, who flitted from group to group, as jovial as the host of a game show, socializing, laughing, and serving beer. His gregarious act was disconcerting, but an integral part of the Bakiga mourning process: celebration in the face of tragedy, a kind of ritualized, group antidote for grief. No one mentioned Hope at all, and I wouldn't learn the details of her death until the following day, when I talked to Karen and Liz.

They told me how she had turned suddenly worse a week ago, and decided to seek treatment at Mulago, the best hospital in the country. Karen helped her and Phenny reach Kampala, navigating the transportation challenge as Hope's health turned quickly from bad to worse. They spent a sleepless night at Liz's house before catching a ride to the hospital. In the car, Hope complained of pain and numbness, first in her feet, and then in her legs and stomach. She lay in the back seat with her head on Phenny's lap, and as her vision started to fade, she raised her hands weakly and waved. "Bye Phenny," she called out as she died. "Bye, Phenny, bye." He cried and they sped through traffic, but it was too late and Hope was gone. Phenny held her eyes shut all the way to Mulago.

The vigil dragged on and I kept myself awake, sitting by the fire and listening to the stories and laughter. As the *tonto* jugs slowly emptied, men started dropping off to sleep, curling up under benches, or finding room on the crowded floor of the house. Phenny never seemed to tire, but finally, in the hour before dawn, I talked to him alone and his face fell, aging twenty years in an instant.

"I am living in a hell on earth," he told me, his voice small and haunted. "Nothing in this life can bring me happiness again."

Most Bakiga people collect several names during their lifetime, family titles or clan slang, some with inexplicable meanings. In the forest, I worked with trackers like Tibamanya—"they don't know," and Bayenda— "they want." Even Phenny and Hope chose appellations with double meaning; Amanda's middle name had been Najebale, "and thank you." As patriarch, Tibasigwa had final say over all the names of his family, and in Phenny he showed a sad talent for prophecy. Since early childhood, he had referred to his son as Vuma, "the suffering one."

"Tour...you...friend." Ephraim was a little drunk. He and Philman, and a few members of the trail crew sat with me on the lawn, watching mist spiral up from the forest and sipping banana beer, the left-overs from my farewell party. Over the past few days I had made speeches, said countless goodbyes, and held graduation ceremonies for the guides in training. The wardens thanked me with a party and a T-shirt, hand painted with a simple epigraph: "We loved you in Bwindi Park."

As my time in Buhoma shrank from days into hours, the poignancy of departure tainted every action with a sense of finality and significance. Familiar faces, conversations, or walks in the forest took on a distant aspect, like old newsreel footage gone grey with memory. The *last* trip to Butagota market; the *last* day with the gorillas; the *last* time safari ants attacked me in my sleep. I felt detached, my sadness at leaving Uganda mixed with a paradoxical relief, and the strong desire to see my home and family.

On the *last* morning, I hiked to the waterfalls alone, reading farewells in every birdcall, and the reverent hush of the forest. Crossing Munyaga River, I felt the bridge beneath me, some poles firm and some gone loose with rot. I remembered lifting the span into place, and how the trail crew laughed when we dropped a corner on Stanley's foot. *"Guma, guma!!"* they chanted for strength as they hoisted the timbers again, and he soaked his foot for an hour in the cold, clear water. *That was years ago*, I thought, and moved on.

My bags packed and my house empty, I loaded everything in Liz's truck. It was a day like any other when I stopped by the office, with tourists visiting both gorilla groups, and several guides out leading forest walks. Phenny and William were there, and they walked me up to the road. I shook hands with William and he wished me well, promising to write. Then Phenny was left.

"I hope you will come back some time," he said hollowly, looking at the ground. "But I don't think I will still be alive."

"Ah, Phenny…" I was at a loss. After Amanda's death, and then Hope, we never spoke of AIDS. But I always knew it, and so did he.

"I am weak," he blurted, sobbing suddenly and moving away. "I am already weak."

I ran a few steps after him and held his shoulders. "I'll see you again, Phenny," I said fiercely, to myself as much as to him. "I'll see you again."

He wept and walked off without answering, and I crouched down in the road, swallowing tears. Behind me, I heard Liz start the truck's engine and switch on some music. My last glimpse of the forest was a blur of green as I turned, climbed into the cab, and drove away.

((((

Fishermen waded like tall storks through the shallows, casting their cinch nets with practiced ease and pulling them in hand over hand, faint ripples in blue water. Further out, I watched a group of women gathering seaweed from the reef and chasing schools of silver minnows towards the bamboo poles of a fish trap. In the evening, muezzin cries drifted out from a dozen mosques, like mad auctioneers calling the faithful to prayer.

Under a sunset sky of liquid orange, I wandered through Zanzibar's fabled port district, Stone Town, one of the oldest cities in East Africa. Ancient whitewashed walls rose up around me and narrow alleys branched in every direction, a labyrinth of shady courtyards, fish markets, and spice shops. Occasional motor scooters careened past and I could hear traffic on the main streets, but much of the town hadn't changed in centuries, a tranquil, Arabic place where cats lounged in elaborate doorways of hand-carved wood, studded with polished brass. Zanzibar took me to another Africa, far from my world in Bwindi, an exotic pause between the chaos of departure, and the shock of going home.

I travelled east across the island through a countryside redolent with spices: clove trees, cardamom, cinnamon, and nutmeg. Combined with the trade in slaves and ivory, these crops gave Zanzibar an age-old link to the Middle East and India. Huge wooden dhows plied the trade routes for centuries, giving rise to Swahili, the distinctive culture of Africa's eastern coast. Blending local customs with the lifestyles of India and the Arabian peninsula, Swahili people lived in a cosmopolitan world long before traders and explorers penetrated inland. The Portuguese added a European influence in the sixteenth century, but lost hold of the region to the powerful Sultan of Oman, who eventually moved his capital to Zanzibar. When Europeans returned in the mid-1800s, they found a thriving island city that served as the starting and end points for countless African journeys, from the time of Speke and Stanley well into the twentieth century.

For me, Zanzibar was a refuge. The conclusion of my time in Uganda left me with a strange mixture of release and regret, complicated by the tragedy of Hope's death and the wrenching stress of parting with friends. I needed a transitional landscape, a foreign, anonymous place where I could restore myself before moving back home. After a lifetime of travel and exploration, Henry Morton Stanley described the subtle disappointment of a journey's end:

"When a man returns home and finds for the moment nothing to struggle against, the vast resolve, which has sustained him through a long and difficult enterprise, dies away, burning as it sinks in the heart; and thus the greatest successes are often accompanied by a peculiar melancholy."

When I left the Impenetrable Forest, my years there became instantly distinct, a finite memory. I felt Stanley's 'peculiar melancholy,' but only briefly. It faded with my weariness as the weeks passed on Zanzibar, and I found a new sense of purpose, another challenge to struggle against.

In the village of Bweju, there was a small shop selling school supplies: cheap ball point pens, and undersized notebooks decorated with cartoon dinosaurs and Arabic script. I bought one of each, and returned to my favorite stretch of sand by the ocean. Overhead, lilac-breasted rollers screeched and fluttered in the dry fronds of a palm tree, and two sooty gulls soared on the trade winds, their wings curved backwards like twin parentheses over the blue-green waves of the reef. Picturing Bwindi, and my last day with Katendegyere group, I uncapped the pen, opened the book, and began to write...

We hiked through a heavy grey downpour, up over Rushura and across the border into Zaire. Water pooled brown in our boot prints as we crossed a fresh-turned field, the dark earth heaped in mounds for sweet potato and yams. I walked in front with Charles, and he chose a path leading sharply down through dense thickets, to a patch of forest outside the park. The gorillas had been in this area for several days, making small raids in a neighboring banana plantation.

We found them quickly, hunkered down in a bower of shrubs to avoid the rain. I saw a shoulder, and part of a broad back that might have been Karema, but leaves and mist obscured the rest. After nearly an hour, the storm began to lift and heavy sunbeams fell down through the breaking clouds. I listened to a sharp cough-grunt as Mugurusi came to life, then the snapping branches of movement. We followed slowly, but the undergrowth thickened, and I thought I would have to leave Bwindi without another clear sight of the group.

Suddenly, the forest gave way to open space, a new field, recently cleared and burned for millet. All of Katendegyere group were there, black shadows moving cautiously through the damp ash. Kasigazi followed close behind his mother while Katome, Karema, and Kacupira brought up the rear, pausing to tear wet leaves from a tree at the edge of the forest. Mugurusi turned back towards us and stopped. His long hair was matted with rain, and it shone like dull pewter down the length of his back. He stood there for several minutes, regarding us calmly while his family crossed the blackened field behind him and disappeared, one by one, into a wall of green.

EPILOGUE

*"Massacre in the Mist"... "Death March"... "In Uganda,
Vacation Dreams Turn to Nightmares"*
—Headlines from *U.S. News & World Report,
Newsweek,* and *Time Magazine, 1999*

Phenny awoke to the sound of gunfire. He'd heard the rumors of Hutu militia in the area, and had slept the past few nights with a loaded park rifle tucked under the bed. Grabbing the gun, he ran out into the early light of a Buhoma morning and began shooting to protect the park, to protect his home, to protect his new wife and their young child.

Already the campground was in flames and deafening grenade explosions announced the end of the park's new vehicles. Phenny saw dozens of ragged soldiers descending the path from Rushura Hill and emptied his rifle in their direction. He watched one man fall before throwing down the useless gun and fleeing into the forest, the bullets of their return fire hissing through the vegetation around him.

Phenny and his family survived the attack, but by the end of that day nine people would lie murdered in Bwindi Forest. Community Conservation Warden Paul Wagaba died with his hands bound, doused with gasoline and burned alive by his Rwandan captors. The rebels then singled out English-speaking tourists and led them away up the Muzabajiro Loop trail, where eight Britons, Americans and New Zealanders were hacked to death with *pangas*.

Stories from survivors and notes pinned to the bodies explained the attack as a calculated political move. Thousands of Rwandan Hutus, the same people responsible for the genocide of up to a million of their Tutsi countrymen in 1994, had lived in exile in Zaire since losing power in Rwanda's civil war. They targeted the tourism program at Bwindi to send a message to the world that their struggle for power was still alive, and to punish Uganda, Britain and the United States for their support of the new Tutsi-led government in Rwanda.

The massacre brought tourism in Uganda to a standstill and showed how vulnerable development and conservation efforts are to the violence of political unrest. The fate of Uganda's people and the fate of its nature are inextricably linked. "They have closed the park," Tibesigwa Gongo lamented to me in a letter. "We have nothing to chew."

Journalists spotlighted the murders as an inherent risk of adventure travel, but missed the underlying tragedy, the fact that Tibesigwa, Phenny and everyone else in Buhoma do not have a safe home to return to. This attack occurred in the heart of their community, and they must live with the constant threat of further violence, while watching the disruption and decline of a promising tourism industry on which they had pinned so many hopes and aspirations.

I received the news with numb shock, venting my horror in frantic, vain attempts to contact friends in Uganda and get accurate information. Was everyone all right? What would happen now? Suddenly, the people, sights, smells and other memories of life in the Impenetrable Forest were vivid in my mind again. In the years since returning from Uganda I had come to think of my time there as something historical, a life within a life. Letters from Tibasigwa, Twinomujuni and others had kept me abreast of the changes in Bwindi as the world I knew there slipped slowly into the past.

Soon after my departure, Katendegyere group had begun to fall apart. Katome and Karema dispersed to travel alone, and Mugurusi passed away from natural causes, leaving Kacupira as the lead silverback. He still travels with Nyabutono and Kasigazi, but they shifted their range to the steep hills east of Munyaga River. Trackers and tourists are sometimes forced to drive several miles around the edge of the forest before starting their hike, and park officials plan to stop visiting them altogether as soon as an alternate group can be habituated.

James Mishana, the stalwart Katendegyere tracker, died after a short illness, but the rest of the staff is healthy and still involved with parks and tourism. Medad Tumagabirwe went on to study wildlife management in Queen Elizabeth Park, while Alfred Twinomujuni has become an expert on rainforest birds, leading walks for visiting ornithology tours. Phenny Gongo received a promotion and now supervises all of Bwindi's gorilla habituation activities, and Betunga William led a group of visitors that became the first people to witness a gorilla birth in the wild. The Community Campground has continued to flourish, and recently purchased a pick-up truck to provide regular transport to markets and hospitals.

With no direct contact for months after the massacre, I had to imagine Buhoma's response at every level. Thinking it through, I realized that while the details may have changed, the habits and customs of daily life were probably just as I'd known them. It was in those fundamentals that I found hope. For a nation and a people that survived Colonialism, Idi Amin and Milton Obote with their humor and kindness intact, a setback like the Hutu attack on Bwindi was a bump in the road. The park would reopen, tourists would return, and people would rebuild their lives and dreams as they always had.

For Bwindi's mountain gorillas, survival will forever be the struggle of a fragile population surrounded by threats of encroachment. Armed rangers continue to patrol near the habituated groups and there has been no poaching since the Kyaguliro incident, but Bwindi is a limited island of habitat and the gorillas will always be rare creatures. Tourism has created an economic incentive for conservation at the local and national level, and that ethic should continue to grow as the program rebuilds and moves forward. If the people of Uganda can maintain political stability and find hope for the future, then there must be hope for the gorillas and forests as well.

The guides now tell me that Ruhondeza, Kacupira, and the other silverbacks have learned to fear Makale above all others, steering their families away whenever he comes near. When he and Mutesi finally succeeded in luring a female away from Mubare group, Makale quickly chased off his older brother and kept her for himself. The pair have been sighted on the slopes of Rukubira, Rushura and along Muzabajiro creek, a range that overlaps almost exactly the former home of the Katendegyere family. "Any time now, they should produce a kid," writes Alfred. We'll have to wait and see.

GLOSSARY

The Bantu family of African languages includes hundreds of dialects throughout the sub-Saharan region. Rukiga is spoken only in the south-western part of Uganda, but is structured the same way as all Bantu languages and is very similar to the dialects of neighboring tribes. The language contains several distinct noun classes, each of which have different prefixes to denote plural and singular forms, and sometimes the size of the object. People fall into the *Mu-Ba* noun class, so a single white person is a *muzungu*, while many are *bazungu*. Similarly, a single tribe member is a *mukiga*, while several, or the tribe as a whole are referred to as *Bakiga*. Greetings often include a long series of questions and answers, starting with general news and leading to specific queries about the health of one's family and crops. Only the simple forms, and other words used in the book, are included here. Verb stems are listed alphabetically, with their infinitive prefix (*ku*) in parenthesis.

Luganda and Swahili are also bantu-based tongues, although many of the noun prefixes are different, and Swahili includes Arabic influences stemming from its origins in the trading culture along the Kenyan and Tanzanian coast. Words from these languages are denoted with a capital L or S respectively.

Abantu	-	People
Agandi	-	What's the news; how are you.
Amabare	-	Stones.
Bahima	-	Ruling class of Ankole kingdom.
Banda	-	(S) traditional house, usually round, with thatch roof.
BaaNyabo	-	Ladies.
BaaSsebo	-	Sirs; gentlemen.
Dudu	-	(S) Insect, bug.
Dhow	-	(S) Type of sailing vessel on E. African coast, introduced by Arab traders.
Ebihimba	-	Beans.
Ebitakuri	-	Sweet potatoes.
Eego	-	Yes.
Eki	-	What.
Emondi	-	Potatoes.
Empazi	-	Safari ants.
Engagi	-	Gorilla.
Enjojo	-	Elephant.
Enjoka	-	Snake.
Enkima	-	Monkey.
Enkobe	-	Baboon.
Erizooba	-	Today.
(ku)garuka	-	to go back.
Goma	-	(L) Traditional dress for women. Brightly colored satin with a wide sash and puffed shoulders.
(ku)gyende	-	to go.
Hamwe	-	Together.
Iwe	-	You (singular).
Jebale	-	(L) Thanks for working.
Jjuko	-	(L) An Mborogoma clan name.
Kabaragara	-	Small, sweet bananas.

Ka	-	Let, as in *Ka Tugyende*: "Let's go."
Kanju	-	(L) Traditional dress for men, a long white smock.
Kare	-	O.K. -sometimes used as a greeting.
Karoti	-	Carrots.
Kashundwe	-	Wart.
Ki	-	What.
Kodi	-	Greeting—"Is anyone home?"
(kw)ija	-	to come.
Makyita	-	Courage/mischief.
Matatu	-	Taxi van or pickup; also a cardgame.
Matoke	-	Green bananas, served steamed and mashed.
Mborogoma	-	(L) Lion, a clan totem.
Mbuzi	-	Goat.
Mpora	-	Sorry.
Mpora mpora	-	Slowly by slowly.
Muchuchu	-	Shadow; dusty.
Munonga	-	A lot; very much.
Munyaga	-	Thief; robber. Name of river in Bwindi Forest.
Munwani	-	Friend.
Muraaregye	-	Spend the night well; goodnight (to several people)
Muzee	-	Old man/sir.
Mwakora	-	You have worked; thanks for the work. A typical greeting for people digging in the fields or doing other labor.
Na	-	And.
Ndi gye	-	I'm fine; it's good.
Ngaha	-	No.
-ngahi	-	How much; how many.
Nooseera	-	You're overcharging me.

Nyabo	-	Madam.
Obugyeni	-	Party; celebration.
Oburo	-	Millet.
Omubisi	-	Sweet banana juice; unfermented *tonto*.
Omwifa	-	A common forest tree and gorilla food (*Myrianthus* spp.)
Oraaregye	-	Spend the night well; goodnight (to one person).
posho	-	Corn meal porridge (boil water; stir in white corn meal until you can't move the spoon; eat)
(ku)ronda	-	to look for.
Rugabo	-	Silverback.
(ku) reeba	-	to see.
(ku) ruhuka	-	to rest.
Shamba	-	Small farm.
Ssebo	-	Sir.
(ku) teemba	-	to climb.
Tonto	-	Banana beer.
Tu-	-	We.
Sula burungi	-	(L) Sleep well; goodnight.
Waragi	-	Local banana whiskey; also a brand name for Ugandan gin.
Webale	-	Thank you

Suggested Reading

Akely, Carl. *In Brightest Africa*. New York: Garden City Publishing Company, Inc., 1920.

Akely, Mary L. Jobe. *Carl Akely's Africa*. New York: Dodd, Mead and Company, 1930.

Ballantyne, R. M. *The Gorilla Hunters*. Philadelphia: Henry T. Coates & Co., c. 1890.

Baumgartel, Walter. *Up Among the Gorillas*. New York: Hawthorn Books, Inc., 1976.

Burton, Richard Francis. *The Lake Regions of Central Africa*. New York: Harper, 1860.

Carey, M. and E. H. Warmington. *The Ancient Explorers*. New York: Dodd, Mead & Company, 1929.

Cook, David & David Rubadiri, editors. *Poems from East Africa*. Nairobi: East African Educational Publishers Ltd., 1971.

DuChailu, Paul. *Wildlife Under the Equator*. New York: Harper & Brothers Publishers, 1869.

Edel, May M. *The Chiga of Uganda, Second Edition*. New Brunswick: Transaction Publishers, 1996 (1st edition 1957).

Fossey, Dian. *Gorillas in the Mist*. Boston: Houghton Mifflin Company, 1983.

Goodall, Jane. *In the Shadow of Man*. New York: Dell Publishing Company, 1971.

Hall, Richard. *Lovers on the Nile: The Incredible African Journeys of Sam and Florence Baker*. New York: Random House, 1980.

Hansen, Bernt Holger and Michael Twaddle, editors. *Changing Uganda*. London: James Currey Ltd., 1991.

Hansen, Bernt Holger and Michael Twaddle, editors. *Uganda Now*. London: James Currey Ltd., 1988.

Hemingway, Ernest. *Green Hills of Africa*. New York: Charles Scribner's Sons, 1935.

Johnson, Martin. *Congorilla*. New York: Harcourt, Brace and Company, 1931.

Kyembe, Henry. *State of Blood: The Inside Story of Idi Amin*. New York: Ace Books, 1977.

Lamb, David. *The Africans*. New York: Vintage Books, 1983.

Moore, Gerald and Ulli Beier. *The Penguin Book of Modern African Poetry*. New York: Penguin Books, 1984.

Moorehead, Alan. *No Room in the Ark*. New York: Harper & Brothers Publishers, 1957.

Mutibwa, Phares. *Uganda Since Independence: A Story of Unfulfilled Hopes*. Trenton: African World Press, Inc., 1992.

Ngologoza, Paul. *Kigezi and its People*. Kampala: East African Literature Bureau, 1969.

Portal, Sir Gerald. *The British Mission to Uganda in 1893*. London: Edward Arnold, 1894.

Roscoe, John. *The Soul of Central Africa*. London: Cassell and Company, Ltd., 1922.

Schaller, George B. *The Mountain Gorilla*. Chicago: University of Chicago Press, 1963.

Schaller, George B. *The Year of the Gorilla*. New York: Ballantine Books, 1964.

Speke, John Hanning. *Journal of the Discovery of the Source of the Nile*. New York: Dutton, 1969 (reprint; original 1864).

Stanley, Henry Morton. *In Darkest Africa*. New York: Charles Scribner's Sons, 1890.

Stanley, Henry Morton. *The Autobiography of Sir Henry Morton Stanley*. Boston: Houghton Mifflin Company, 1909.

Stanley, Henry Morton. *Through the Dark Continent.* New York: Harper and Brothers, 1879.

Taylor, Bayard. *The Lake Regions of Central Africa.* New York: Charles Scribner's Sons, 1881.

Wasserman, Jacob. *Bula Matari—Stanley, Conqueror of a Continent.* New York: Liveright Publishers, Inc., 1933.